PHYSICS

What's the problem?

R. H. C. Neill

G. Sydserff

Edward Arnold

© R.H.C. Neill, G. Sydserff 1978

First published 1978
by Edward Arnold (Publishers) Ltd
41 Bedford Square, London WC1B 3DQ

Reprinted 1981

ISBN 0 7131 0201 2

British Library Cataloguing in Publication Data

Neill, R H C
    Physics.
    1. Physics – Problems, exercises, etc.
    I. Title   II. Sydserff, G
    530'.076   QC32

ISBN 0 7131 0201 2

Set in 11pt IBM Press Roman by Preface Ltd., Salisbury, Wilts.
Printed in Great Britain by Spottiswoode Ballantyne Limited,
Colchester and London

Authors' preface

The authors feel that many pupils find considerable difficulty in relating the problems which they are required to solve to the underlying theory. With a packed syllabus to cover, the teachers, however willing, cannot always find the time to develop the pupils' examination techniques as well as they would like to. It is hoped that 'Physics — What's the Problem?' will go some way towards overcoming these difficulties.

The book is presented in such a way that it can be used by pupils who want an effective method of revising Ordinary Grade/Level Physics for examinations, or by pupils who want to improve their problem-solving techniques. In the introductory section on 'How to Use this Book', we have stressed that the book has been structured to cater for different individual needs. There are five units in the text covering the principal subject areas encountered in most O-Grade/Level courses (Waves, Mechanics, Heat and the Gas Laws, Electricity & Radioactivity). Within each Unit, the topics are covered by a sequence of examples to which **fully worked solutions are provided**. The examples and their solutions are printed in separate sections of the book so that the reader can study the material in a variety of ways.

S.I. Units have been used throughout, and all numerical calculations have been kept simple so that the physics is not obscured by a maze of complex arithmetic. The Worked Examples have been given a mark allocation but this is only included for guidance.

We hope that 'Physics — What's the Problem?' will appeal to teachers searching for problems, and to pupils searching for answers.

Robert H. C. Neill

George Sydserff

Edinburgh, December, 1977.

Acknowledgments

We are indebted to our artist, Tony Merriman, for his excellent interpretation of our own, very basic sketches. His professional involvement in teaching physics and his irrepressible sense of humour are apparent in his illustrations and diagrams.

Contents

How to use this book

How this book is set out

There are five UNITS in this book, covering the five major topics of most Ordinary Grade/Level Physics courses:

unit W on **Wave motion**
unit M on **Mechanics**
unit H on **Heat and the gas laws**
unit E on **Electricity**
unit R on **Radioactivity**

Each Unit has two main parts:

(a) A set of questions presented in an acceptable teaching order.
(b) A corresponding set of detailed solutions to these questions.

In addition, each Unit has a batch of practice problems which are modelled on the above worked examples. Answers only are supplied for these practice problems and these are given at the end of the book.

There are several ways of using this book

- If you want a way to revise your physics course **quickly** for an examination, read each example in conjunction with its solution and work your way steadily through the Units.
- If you want to practice solving physics questions try to do each question yourself without looking at its solution. Then check your answer carefully against the solution provided.
- If you want to study questions on a particular topic (e.g. Ohm's Law) look up this topic in the index at the back of the book to find a list of all the relevant examples. Then study these carefully.

Remember to try the practice problems. They will test your progress.

UNIT W

Wave motion

Content List of Topic Areas

Worked Examples

W1

Introducing Dave, dog and travelling waves.

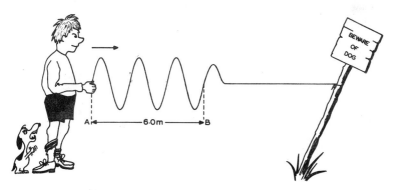

2 Wave motion

(a) Are the waves which Dave produces longitudinal or transverse? (1)
(b) Describe how he produces this wave motion in the rope (2)
(c) What name is given to the horizontal distance between one crest and the next crest? (1)
(d) What would this distance be if AB is 6.0 metres? (1)
(e) Dave makes 10 complete, up-and-down movements of his arm in 5.0 seconds. At what speed do the waves travel along the rope? (2)
(f) How long would it take one wave to travel the distance AB? (2)
(g) What would happen at the fixed end (the post) when the waves arrive? (1)

W2

As you can see, physicists can be sportsmen too! This time, the rugby posts and the crossbar have come in useful for an experiment on wave motion. A long coil spring (or 'Slinky') is suspended horizontally by a number of strings attached to the crossbar.

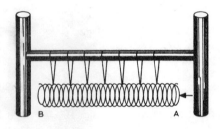

(a) If the coils at end A are given a short, sharp push in the direction shown, what will happen? (2)
(b) End B is now fixed to the upright post. Describe what happens when end A is given another short, sharp push. (1)
(c) What kind of wave motion, longitudinal or transverse, is being produced in (a) and (b)? (1)
(d) How would you adapt the result observed in (a) to describe what happens when a long steel rod is hit, end on, with a hammer? (3)
(e) How would you measure the speed at which a pulse travels along the spring? (3)

W3

These two diagrams represent cross-sections, or profiles, of water waves produced by a plane wave vibrator in a ripple tank.
A small fishing float is shown on the water surface at O in diagram I. It is shown, 0.1 second later, in position O' in diagram II. During this time interval, the wave moves to the right and the float drops to a position 2.0 cm below position O.

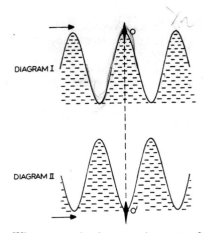

(a) What name is given to the part of the wave near to O in diagram I? (1)
(b) What name is given to the part of the wave near to O′ in diagram II? (1)
(c) If O and O′ are the highest and lowest positions of the float, what is the wave amplitude in the region near the float? (1)
(d) Calculate the frequency of the plane wave vibrator in Hz. (2)
(e) If the waves travel across the water at 20 cm s^{-1}, what is their wavelength? (2)
(f) Energy will be needed to raise the cork up to O again. Where will this energy come from? (2)
(g) Would the motion of the float be the same if it were repositioned further away from the source of the waves? (1)

W4
A point dipper P is moved in and out of the water in a ripple tank at a regular frequency, producing circular waves as shown. A flexible metal strip ABC acts as a barrier to the waves.

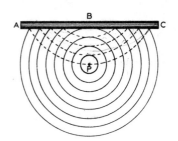

All of the waves shown in the above diagram were produced by the dipper in 0.5 s. The metal barrier is 0.02 m away from P.
(a) Calculate (i) the wavelength,
(ii) the frequency,
(iii) the speed of the waves. (3)

(b) What do the broken lines in the diagram represent? (1)

(c) What happens to (i) the speed, and (ii) the frequency of the waves after they strike the barrier? (2)

(d) What device could you use to produce an apparently stationary pattern like that shown in the diagram? Describe briefly how you would use the device you have chosen. (2)

(e) If ends A and C of the barrier are pulled down to make it curved, how would this affect the wave reflecting from the barrier? (2)

W5

(a) This is a pupil's eye view of a ripple tank. The vibrator is sending plane waves across the tank from left to right.

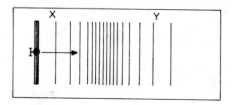

(i) What is happening to the waves over region XY? (2)

(ii) Explain how this effect is produced. A diagram would assist your answer. (3)

(b) In the ripple tank below, there are two wave generators, A and B. A is in the deeper water to the left of the line XY and B is in the shallower water to the right of XY. The generators produce waves of the same frequency.

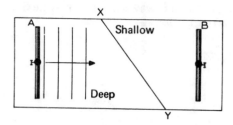

(i) Draw a diagram of the waves produced in the tank when A is operating on its own. (3)

(ii) As in (i), but with B operating on its own. (2)

W6

(a) Copy and complete the diagrams shown. Each one represents plane waves encountering a barrier.

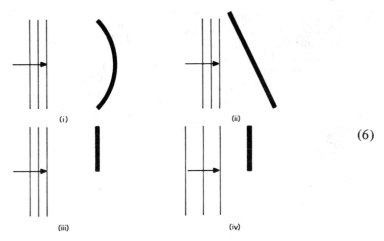

(i)   (ii)

(6)

(iii)   (iv)

(b) In the study of water waves, what is meant by the terms constructive and destructive interference? Explain how each comes about. (2)

(c) A regular interference pattern is to be produced in a ripple tank by using two point dippers. Describe how you would set them up. (2)

W7

The following diagrams illustrate how plane water waves change shape when they pass through gaps in barriers.

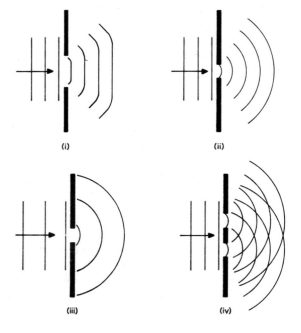

(i)   (ii)

(iii)   (iv)

6 Wave motion

(a) What is this bending effect called?   (2)
(b) What do the diagrams indicate
    (i) about the relationship between the gap width and the
        amount of bending?
    (ii) about the relationship between the wavelength and the
        amount of bending?   (4)
(c) A small barrier inserted into the wide gap of diagram (i)
    produces the wave pattern shown in diagram (iv). What
    other property of waves is now evident in the region where
    the waves overlap? Explain how such an effect comes
    about.   (4)

W8
(a) Assuming that light has wave properties, what property of the
    wave would determine
    (i)   the colour,
    (ii) the brightness, or intensity, of the light?   (2)
(b) Explain why red light shows more diffraction than blue
    light when each is passed in turn through a very narrow
    slit.   (2)
(c) The diagram below represents an interference effect that can
    be readily produced with light waves.

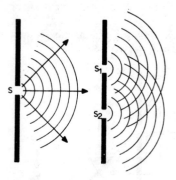

(i)   Suggest a suitable source of light for this experiment.   (1)
(ii) How would you construct the double slit arrangement
    S_1 and S_2?   (2)
(iii) How would you view the interference effect?   (1)
(iv) Describe briefly what you would expect to see.   (2)

W9
Each of the following diagrams shows a ray (or rays) of white light
entering boxes which have glass windows on two opposite sides.
Looking from above, it is impossible to see what is inside each one.
The contents of the boxes affect the light in the following ways:

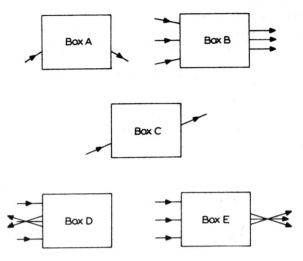

Copy and complete the diagrams to show the contents of each box
and the effect of the contents on the light. (5 x 2)

W10
A pupil modifies a ray box by surrounding it with a rigid structure of
three colour filters and a piece of clear glass. She then places it, along
with a lens, in front of a glass prism. The ray diagram obtained with
the clear glass transmitting the light is shown below.

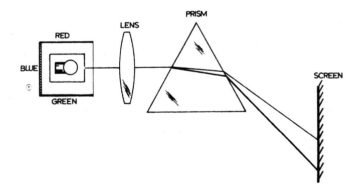

(a) What is the function of the lens? (1)
(b) What name is used to describe what happens to the light
when it passes from air into glass and changes direction? (1)
(c) Describe what the pupil will see on the screen when the
clear glass is in front of the ray box slit. (2)
(d) What will she see on the white screen when each colour
filter is placed, in turn, in front of the ray box slit? (2)

(e)   She then demonstrates to her friend that infra-red
radiation is falling on the screen when the clear glass is
in position.
   (i)   What kind of detector could she use to detect the
         infra-red radiation?   (2)
   (ii) Where should she place the detector to pick up an
         infra-red signal near the screen?   (2)

W11

In the diagram, T represents a short wave radio transmitter, often
referred to as a microwave transmitter. R is a receiver which is
connected to an amplifier and loudspeaker unit. The transmitter is
positioned in front of a reflector as shown.

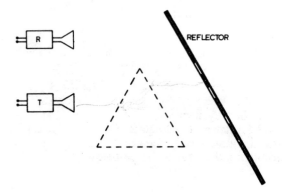

(a)   What material would make a suitable reflector?   (1)
(b)   Draw a diagram to show where the receiver should be
      placed to obtain the maximum reflected signal.   (2)
(c)   A prism filled with paraffin is placed over the dotted outline and
      the receiver is then repositioned for maximum reception. Redraw
      the diagram to show the path taken by the microwaves and to
      show where the receiver should now be placed.   (3)
(d)   If other pieces of suitable reflector are available, show how you
      would demonstrate to a friend that wave interference is possible
      with microwaves.   (4)

W12

Our roving, lunar radar station consists of a transmitter T, a receiver
R and a monitoring oscilloscope. The transmitter sends out regular
pulses of high frequency radio waves and these are reflected back
from distant objects to the receiver. The monitor is an oscilloscope
which displays the received pulses on the screen. In practice, the
objects are so far away, compared with the distance separating T and

R, that the reflected waves reaching the receiver travel in almost the same straight line as the outgoing waves.

(a) The frequency of the transmitter is 1.5×10^{10} Hz and the radio waves travel at 3.0×10^8 m s^{-1}. Calculate the wavelength of the waves. (2)

(b) If the time lapse between the transmission of a pulse and the reception of its reflection from a stationary object is 3.0×10^{-4} s, how far away is the target? (2)

(c) Would the rarer atmosphere on the Moon have much effect on the speed of the radio waves? (2)

(d) Why would targets made of metal give traces of greater amplitude on the monitor screen? (2)

(e) If the lunar 'buggy' is stationary and the time between sending out a pulse and receiving its echo is decreasing, what does this tell you about the target? (2)

W13

A robot on planet X walks slowly back from a vertical rock wall clapping his 'hands' once every two seconds.

When he is 80 metres from the wall, his recording system registers sound echoes exactly mid-way between claps.

(a) (i) At what frequency will echoes reach the robot's sensors?   (1)
    (ii) Use the above information to calculate the speed of sound on planet X.   (3)
    (iii) What does this value for the speed of sound suggest about the atmosphere on planet X?   (2)

(b) A second robot seated on top of the rock wall 'hears' each clap 1.25 s after it is produced.
    (i) How far apart are the robots?   (2)
    (ii) How high is the rock wall?   (2)

W14

A tuning fork T, attached to a box amplifier B is set vibrating and it produces sound waves in the air. The waves are picked up by a microphone M, which is connected to an oscilloscope as shown. The air between B and M has been represented in diagrammatic form.

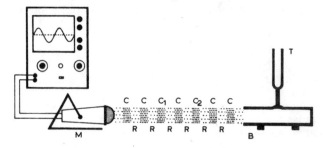

(a) What advantage is there in attaching the tuning fork to a wooden box, B, which is open at one end?   (1)
(b) What do the letters C and R on the diagram stand for?   (2)
(c) What kind of waves are sound waves?   (1)
(d) If the frequency of the tuning fork is 600 Hz, calculate

(i) the wavelength of these sound waves. (1)
(ii) the distance between C_1 and C_2. (1)
(assume that the speed of sound in air is 330 m s^{-1})
(e) Draw diagrams to show how the trace on the oscilloscope
would change in each of the following cases:
(i) T is taken farther away from M
(ii) The wooden box B is removed from T
(iii) A tuning fork of frequency 1200 Hz is used instead. (4)

W15

(a) Morag is asked by her teacher to set up an experiment which will
demonstrate that sound waves can be refracted. The apparatus
at her disposal is shown below.

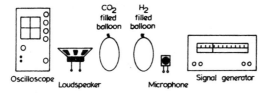

(i) Draw a diagram to show how Morag could utilise this
apparatus for her demonstration. (3)
(ii) Describe briefly how the apparatus could be used to
compare the refraction caused by balloons filled with
carbon dioxide and hydrogen respectively. (2)
(b) Meanwhile, her friend Liz sets up the following apparatus in the
laboratory.

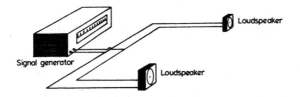

(i) What is this apparatus used to demonstrate? (1)
(ii) Why is it necessary to have two loudspeakers, but only one
signal generator? (2)
(iii) If Liz walks past the front of the loudspeakers, what would
you expect her to hear? (2)

Solutions to Worked Examples

W1

(a) He produces transverse, travelling waves. The particles of the
rope move in the vertical direction as the wave travels
horizontally.

(b) He supplies energy to the rope by moving the free end up and down at a regular rate.

(c) The wavelength, usually denoted λ.

(d) Three complete waves fit into the distance AB and so the wavelength is given by

$$\lambda = \frac{6.0}{3} \text{ m} = 2.0 \text{ m}.$$

(e) There are 2.0 waves per second and so the wave speed is 4.0 m s^{-1}.

Remember the wave equation: $v = f \times \lambda = 2.0 \times 2.0 = 4.0$ m s^{-1}.

(f) Since
$$\text{speed} = \frac{\text{distance}}{\text{time}},$$

it follows that

$$\text{time} = \frac{\text{distance}}{\text{speed}} = \frac{\text{AB}}{v}$$

$$\therefore \quad t = \frac{6.0}{4.0} = 1.5 \text{ seconds.}$$

(g) Waves would reflect at the post and travel back along the rope towards the source of the waves. A stationary, or standing wave pattern, could be set up due to interference between the incident and reflected waves.

W2

(a) A compression pulse will travel along the spring from A towards B. Energy is passed from coil to coil until B is reached.

(b) A compression pulse arriving at B would now be reflected back towards A.

(c) Longitudinal wave motion. The individual particles of the coil move back and forth in the same line as the wave travels.

(d) When a steel rod is hit with a hammer, a compression pulse is sent along the rod. The molecules of steel become displaced and exert forces on neighbouring molecules. The energy supplied by the hammer is transmitted along the rod by the steel molecules. The speed is high because of the close contact of the molecules.

(e) With both ends of the spring tied to the posts, a pulse is produced at end A and a stopwatch is started. The time for the pulse to travel along to B and back **several times** (e.g. four times) can be obtained before the pulse 'dies out'. The total path length can be found once the distance between the uprights has been measured.

Example

distance between uprights = 5.5 m

total distance travelled (path length) = 4 x (2 x 5.5) = 44 m

total time taken = 11 s

$$\text{speed of pulse} = \frac{44 \text{ m}}{11 \text{ s}} = 4 \text{ m s}^{-1}$$

W3

(a) a crest
(b) a trough
(c)
$$\text{amplitude} = \frac{\text{(distance from crest to trough)}}{2} = \frac{2.0 \text{ cm}}{2} = 1.0 \text{ cm}$$

(d) The float moves from O to O' in 0.1 s. It would therefore complete one oscillation in 0.2 s. The float makes 5 oscillations each second and so the frequency is 5 Hz.

(e) Since $v = f \times \lambda$, then
$$\lambda = \frac{v}{f} = \frac{20 \text{ cm s}^{-1}}{5 \text{ Hz}} = 4 \text{ cm}.$$

(f) The energy will come from the waves. It is the wave energy travelling out from the source which causes the surface at any point to rise.

(g) As the distance from the source increases, the wave energy decreases. The motion of the float would be similar to the motion at OO', but the amplitude would be smaller.

W4

(a) (i) There are 4 complete wavelengths between P and B
$$\therefore 4\lambda = 0.02 \text{ m} \quad \therefore \lambda = 0.005 \text{ m (or 0.5 cm)}$$
(ii) 8 waves are produced in 0.5 s.

$$\therefore 16 \text{ waves are produced per second} \quad \therefore f = 16 \text{ Hz}.$$

(iii) $v = f \times \lambda = 16 \times 0.005 = 0.08 \text{ m s}^{-1}$ (or 8 cm s^{-1})

(b) Broken lines represent reflected wavefronts.
(c) There is no change in either speed or frequency during reflection.
(d) A stroboscope could be used. If it is set to flash at the same frequency as the vibrator (16 Hz in this case), the wave pattern will be stationary of wavelength 0.5 cm.
(e) The reflected wavefronts would be less curved. With certain shapes of reflector, the wavefronts could even be plane.

W5

(a) (i) The wavelength is reaching a minimum value and then increasing to the original value. This means that the waves are slowing down, maintaining the lower velocity over a short distance and then they are gradually speeding up again.
(ii) A glass plate with the profile shown below could be immersed in the tank.

(b)   (i)

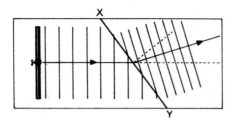

(ii)

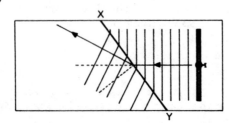

W6
(a)

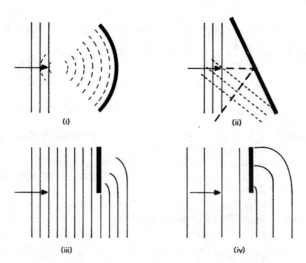

(b)   In regions of constructive interference, the water moves up and
down relatively violently due to the combined effects of the
waves from the two sources. In regions of destructive interference, the water remains relatively calm due to the combined
effects of the waves from the two sources. The explanation of
this is given in the solution to question W7.

(c)   The dippers must move up and down in the water with the same frequency and phase. This can be achieved by suspending the dippers from a rigid spar, which is hung horizontally by two vertical elastic strings or springs. The regular motion is then produced by a motor fixed to the spar as shown.

W7
(a) This effect is known as diffraction.
(b) I. For a given wavelength, the narrower the gap, the greater the amount of bending [compare (i) and (ii)].
II. For a given gap width, the greater the wavelength, the greater the amount of bending [compare (ii) and (iii)].
(c) Interference effects are produced in the region where the waves from the two sources overlap. **Constructive interference** occurs in places where the two wavefronts arrive in phase (i.e. a crest and a crest arrive together, producing a larger crest or a trough and a trough arrive together, producing a larger trough). **Destructive interference** occurs in places where the wavefronts from the two sources arrive out of phase (i.e. a crest from one source arrives at the same time as a trough from the other source). The disturbances tend to cancel out, creating a region of relative calm.

W8
(a) (i) The colour is determined by the frequency of the light.
(ii) The intensity is related to the amplitude of the wave motion.
(b) Red light has a longer wavelength and so it will diffract more than the shorter wavelength blue light.
(c) (i) It is best to use a monochromatic source of light e.g. a sodium lamp which emits light of predominantly one colour.
(ii) Usual method is to scratch two narrow slits with a razor blade on a blackened glass slide.
(iii) The slide could be held close to your eye so that the source is viewed through the two parallel slits or the pattern can be viewed on a white screen.
(iv) The fringe pattern would consist of bright bands of light separated by dark bands, corresponding to the regions of constructive and destructive interference respectively.

W9

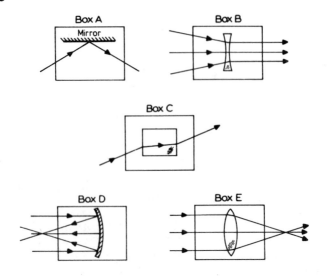

W10

(a)   The lens is used to produce a parallel ray of light from the
      ray box.
(b)   refraction
(c)   **With clear glass**
      A spectrum is produced on the white screen, with the red end
      being refracted least and the violet being refracted most. The
      spectrum colours are red, orange, yellow, green, blue, indigo and
      violet.
(d)   **With red filter**
      Only the red part of the spectrum is transmitted by this filter. The
      refraction in the prism is the same as for the red end of the
      previous spectrum. Approximately one third of the spectrum at
      the red end will appear on the white screen.
      With blue filter
      About one third of the spectrum at the blue end appears on the
      screen.
      With green filter
      Only green light will enter the prism to be refracted. About one
      third of the spectrum, the middle section of the white light
      spectrum, will appear on the screen.
(e)   (i)   She could use heat-sensitive paper, or a phototransistor in
            conjunction with an electrical current meter.
      (ii) The detector should be positioned beyond the red light on
            the screen to obtain the maximum reading, or response.

W11

(a)   A sheet of metal, or metal foil, would make a suitable reflector.

(b)

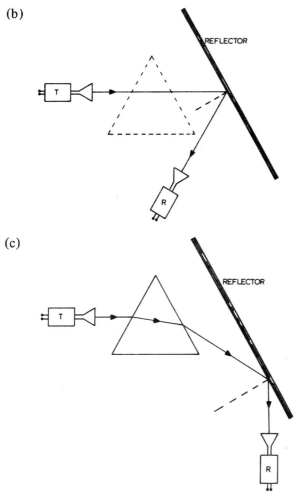

(c)

(d) A slit arrangement could be made as follows using pieces of metal reflector.

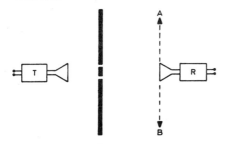

When the receiver is moved along the line AB, the loudspeaker output rises and falls as regions of constructive and destructive interference are encountered. The interference is caused by the overlapping of the two diffracted wavefronts emerging from the slits, whose width should be about the same length as the wavelength.

W12

(a)   From $v = f\lambda$, we get

$$\lambda = \frac{v}{f} = \frac{3.0 \times 10^8}{1.5 \times 10^{10}} = 2.0 \times 10^{-2}\ m$$

(b)   Assuming that the lunar 'buggy' is at rest, the radio waves travel out to the target in 1.5×10^{-4} s. Let the distance required be d metres. Then, $d = v \times t = 3.0 \times 10^8 \times 1.5 \times 10^{-4} = 4.5 \times 10^4$ metres.

(c)   The rarer atmosphere would have virtually no effect upon the speed of the radio waves. Radio waves are electromagnetic and they travel through space at the same speed as light $(3.0 \times 10^8\ m\ s^{-1})$.

(d)   Metal targets act as better reflectors of radio waves so the amplitude of the reflected waves would be greater. This would produce a larger amplitude pulse on the monitor screen.

(e)   It suggests that the target is approaching the 'buggy'.

W13

(a)   (i)   The echoes reach the robot at exactly the frequency at which the claps were produced i.e. 0.5 Hz.

(ii)   Time between clap and its echo $= \frac{1}{2} \times 2 = 1$ s.
Distance travelled by sound in this time
$= 80\ m\ (out) + 80\ m\ (back) = 160\ m$

$$Speed = \frac{160\ m}{1\ s} = 160\ m\ s^{-1}$$

(iii) The atmosphere is not the same as that on earth, where the speed of sound is $330\ m\ s^{-1}$ in normal conditions. It could be some other gas, or gases.

(b)   (i)   Separation of robots = distance travelled by sound in 1.25 s
$= 160\ m\ s^{-1} \times 1.25\ s = 200\ m$

(ii)   Letting H represent the required height of the rock wall, then

$$H^2 = 200^2 - 80^2 \quad \therefore H = \sqrt{33\ 600} = 183\ m.$$

W14

(a)   The wooden box acts as a sound amplifier. The air inside the box is forced to vibrate as the box vibrates. Thus a large volume of air is now vibrating.

(b)   **C** represents a region of compressed air (a compression)
R represents a region of rarefied air (a rarefaction). The pressure in region **C** is greater than normal and that in region **R** is less than normal.

(c) longitudinal waves

(d)
(i) $\lambda = \dfrac{v}{f} = \dfrac{330}{600} = 0.55$ m

(ii) distance $= 2 \times \lambda = 1.10$ m

(e)

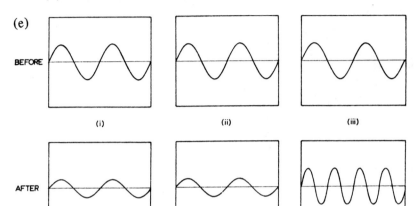

BEFORE

(i)   (ii)   (iii)

AFTER

W15

(a) (i)

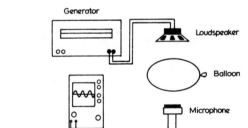

(ii) The balloon filled with carbon dioxide produces a focusing effect which is due to the refraction of sound waves. The action is similar to that produced by a converging lens with light. A strong signal is received by the microphone when it is placed at the region where the sound waves are focused. With the hydrogen-filled balloon there is no such focusing effect because hydrogen is less dense than air and the sound energy is spread out. The action is similar to that produced by a diverging lens with light.

(b) (i) This apparatus is used to demonstrate the interference of sound waves.

(ii) One source is required, with the output divided between the two loudspeakers. It is important that the sound

emitted by the two loudspeakers is in phase and of the
same frequency. The loudspeakers are, therefore, connected in parallel across the signal generator output
terminals.
(iii) She would hear sound interference effects – regions of
maximum and minimum sound intensity. Those are regions of
constructive and destructive interference respectively.

Practice Questions

1. (W1)
A man ties one end of a long rope to a post and then moves the free
end up and down at a frequency of 2.0 Hz. The waves he produces
travel horizontally along the rope towards the post at 3.0 m s^{-1}.

(a) Are the waves transverse or longitudinal?
(b) Use the wave equation to calculate the wavelength of the
     waves produced on the rope.
(c) After the waves hit the post, they reflect back along towards
     the free end.
     What kind of wave pattern is produced by interference between
     the waves travelling to and from the post?

2. (W1)
(a) Write down the wave equation which links the velocity (v),
     the frequency (f) and the wavelength (λ) and give the
     correct units for each quantity.
(b) What is the velocity of waves if their frequency is 50 Hz
     and their wavelength is 0.4 m?
(c) What is the frequency of waves which have a wavelength
     of 5 m and travel at 40 m s^{-1}?
(d) Twenty waves are produced in 5 s on a rope. If the waves
     travel along the rope at 5 m s^{-1}, what is their wavelength?

3. (W2)
A 'Slinky' spring is laid on a smooth bench surface and clamped
at one end.
(a) Describe how a pupil should move the free end of the 'Slinky'
     to produce
     (i)   a transverse pulse,
     (ii) a longitudinal pulse.
(b) A longitudinal pulse is timed as it travels along a 'Slinky'
     and back again **three times**. The time recorded was 18 s
     and the length of the extended 'Slinky' was 6 m. Calculate
     the speed of the longitudinal pulse.

4. (W3)
A piece of cork is floating in a ripple tank. When the vibrator is
operated at 10 Hz, the cork moves up and down as waves of

wavelength 3 cm travel through the water. The distance between the highest and lowest points of the cork's motion is found to be 5 mm.
(a)   What time does the cork take to make one complete oscillation?
(b)   What is the amplitude of the waves as they pass the cork?
(c)   At what speed do the water waves move through the tank?

5. (W5)
Plane water waves travel from deep to shallow in a ripple tank. The direction of the waves changes when they enter the shallow region.
(a)   What name is given to this effect?
(b)   When the waves go into the shallow what happens to
    (i)   the speed?
    (ii)   the wavelength?
    (iii) the frequency?

6. (W5)
When plane water waves travel from shallow to deep water, their speed changes from 20 cm s^{-1} to 30 cm s^{-1}. The waves are produced by a spar vibrating at 5 Hz.
(a)   What is the wavelength in the shallow?
(b)   What is the wavelength in the deep?

7. (W5)
Plane water waves in a ripple tank enter a shallow region whose edge is at 45° to the direction of the incoming waves. Draw a diagram to indicate what happens to the waves due to refraction.

8. (W5)
A ripple tank is tilted so that its base slopes downwards from X to Y.

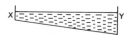

A plane wave generator is operating at end X, at a steady frequency.
(a)   Draw a diagram to show what would be seen, looking down on the tank from above.
(b)   Will the waves travel with a constant speed?
(c)   Will the waves have a constant frequency?

9. (W6)
Plane waves in a ripple tank strike a straight barrier at an angle of 30°
(a)   At what angle to the normal do the waves reflect?
(b)   Due to the reflection what changes (if any) would you expect in
    (i)   wavelength?
    (ii)   velocity?
    (iii) frequency?

10. (W6)

(a)   What could be placed in a ripple tank to reflect plane waves so that they come to a focus? Draw a diagram to illustrate the effect.

(b)   When water waves pass by the edge of a barrier, they bend round behind the barrier to some extent.
  (i)   What name is given to this effect?
  (ii)   Draw a diagram to illustrate the effect.

11. (W6)

Two point dippers are moved in and out of the water in a ripple tank at a regular rate.

(a)   In certain positions, the water surface is calm. Explain this.

(b)   In other positions, large waves are seen on the water. Explain this.

12. (W7)

(a)   How does the width of a gap affect the diffraction of the waves passing through it?

(b)   For a given width of gap, which waves produce more diffraction, low frequency waves or high frequency waves?

(c)   Is there any change in wavelength when waves diffract through a gap?

13. (W7)

Draw a diagram to show what happens when waves of wavelength 4.0 cm pass through two gaps, each 4.0 cm wide, if the gaps are 4.0 cm apart.

14. (W8)

The colours of the spectrum are, in alphabetical order, blue, green, indigo, orange, red, violet, yellow.

(a)   Which colour has the longest wavelength?

(b)   Which colour has the shortest wavelength?

(c)   Which colour would produce interference fringes of the greatest width when passed through a double razor slit?

(d)   Which colour predominates in a sodium vapour lamp?

(e)   What colour would be produced by mixing all the colours of the spectrum together?

15. (W9)

Three parallel rays of light are obtained from a ray box assembly. Draw diagrams to show the effect of placing each of the following in the way of the three parallel rays:

(a)   a concave mirror

(b)   a converging (convex) lens

(c)   a diverging lens

16. (W10)

(a)   Draw a diagram to show how a glass prism can be used to

produce a spectrum of colour from a single ray of white light.
(b) What colour is refracted by the prism
    (i) most?
    (ii) least?
(c) What electromagnetic radiation has a wavelength slightly longer than red light?

17. (W11)
Some properties of radio waves can be studied in the laboratory using microwaves of wavelength 3.0 cm.
(a) What type of material gives the best reflection of microwaves?
(b) What liquid is usually used to show the refraction of microwaves?
(c) What gap width would you choose in an experiment to demonstrate the diffraction of microwaves?
(d) List the apparatus required to show the interference of microwaves.

18. (W12)
(a) Place the following electromagnetic radiations in order of increasing wavelength:
    X-ray, visible light, infra-red radiation, ultra-violet radiation, gamma radiation, radio waves.
(b) At what speed do all the above radiations travel through space (or air)?
(c) Estimate the frequency of the microwaves used in school experiments.
(d) V.H.F. radio waves are transmitted from a station at a frequency of 100 MHz. Estimate their wavelength.

19. (W13)
Some pupils wire up a lamp and a horn to emit a flash of light and a pulse of sound at the same time. They set the equipment up in a wide open space.
(a) Pupils 30 metres away from the unit claim that the flash and 'honk' reach them together whereas pupils 300 metres away claim that they see the light before they hear the 'honk'. Explain this.
(b) When the horn is directed towards a distant building, an echo is received 3 seconds after the 'honk' is emitted. How far away is the building, assuming that the velocity of sound in air is 330 m s^{-1}?

20. (W14)
A vibrating tuning fork emits a note of frequency 660 Hz. The sound waves are detected by using a microphone attached to an oscilloscope.
(a) Are sound waves transverse or longitudinal?
(b) At what speed would you expect the sound waves to move through the air?

(c) Calculate the wavelength of the sound waves.
(d) A wave pattern appears on the oscilloscope screen. What happens to the pattern if
    (i) a tuning fork of lower pitch is used instead,
    (ii) the microphone is moved farther away from the tuning fork?

21. (W15)

Draw a labelled diagram to show how sound waves, obtained from a loudspeaker attached to a signal generator, can be refracted by a balloon filled with carbon dioxide. Show where a microphone connected to an oscilloscope would have to be placed to pick up the strongest signal from the refracted sound waves.

22. (W15)

Make a list of all the equipment you would require to demonstrate interference of sound waves in a laboratory.

UNIT M

Mechanics

Content List of Topic Areas

Worked Examples

M1
A disc, which has a single line painted on it, is rotating
revolutions per second in the direction shown.

Describe how the disc would appear when it is illuminated by
light from a stroboscope, set at each of the following flash-rates:
(a)   10 flashes per second (f.p.s.),
(b)   19 f.p.s.,
(c)   20 f.p.s.,
(d)   21 f.p.s.,
(e)   40 f.p.s.   (10)

M2
(a)   A ticker-timer operates at a frequency of 50 Hz. What
time interval is represented by the section XY of the
ticker-tape shown below?

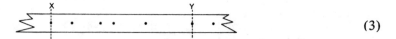

(3)

(b)   A ticker-tape is moved through a ticker-timer for 5.0 s
as measured on a stop-watch. If the timer is operating at
50 Hz, how many dots would you expect to be printed
on the tape?   (3)

(c)   If section OP of the ticker-tape shown below is produced
by a ticker-timer in 0.30 s, at what frequency is it
operating?

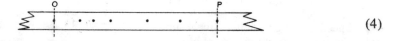

(4)

M3
Rory McBlister takes 1 hour 40 minutes to climb from his base-camp
(X) to the summit of Ben Phew (Y). He travels a total distance of
2400 m up the mountainside and rises 1200 m above X as shown.

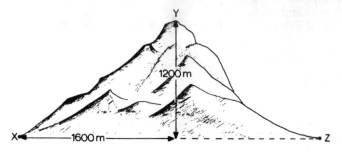

(a) What is his total displacement, in metres, in moving from
     X to Y? (2)
(b) What is his average speed, in m s^{-1}, for this climb? (2)
(c) Calculate his average velocity for this climb. (2)
(d) His friend Jock McCrafty starts out from Z **20 minutes
     after** Rory left X. He takes a much easier route up the
     mountain at an average speed of 0.5 m s^{-1} and reaches
     the summit **20 minutes before** Rory.
     What total distance did Jock travel along the mountain
     from Z to Y? (4)

M4

(a) The table below gives the velocity of a car at intervals
     of one second as it accelerates from rest along a straight
     road.

time (in s)	0	1	2	3	4
velocity (in m s^{-1})	0	4	8	12	16

Calculate the acceleration of the car. (2)
(b) The following table gives the velocity of a marble at
     intervals of 0.2 s as it decelerates up a straight ramp.

time (in s)	0	0.2	0.4	0.6	0.8
velocity (in m s^{-1})	2.4	1.8	1.2	0.6	0

Calculate the deceleration of the marble. (2)
(c) Objects falling freely under gravity in an area where air
     resistance is negligible accelerate at 10 m s^{-2}.
     (i)   What velocity is reached by a ball 0.4 s after it is
           dropped from rest?
     (ii) What velocity is reached by a ball 0.4 s after it is
           thrown downwards at 8.0 m s^{-1}?
     (iii) What velocity is reached by a ball 0.4 s after it is
           projected vertically upwards at 8.0 m s^{-1}? (6)

M5

(a) Giovanni Graviti drops a pebble from the top of the Leaning
     Tower of Pisa and it takes 3.2 s to reach the ground below.
     Assuming that air resistance is negligible,
     (i)   calculate the velocity of the pebble just before it
           strikes the ground, and
     (ii) estimate the vertical distance dropped by the pebble. (4)
(b) Giovanni then drops a much larger stone from the same
     position. Comment upon the time that it takes to reach
     the ground. (2)
(c) Finally, he throws another pebble **vertically upwards** from
     just over the edge of the tower. The following velocity-
     time graph represents the motion of this pebble from the

instant that it leaves his hand to the instant that it strikes
the ground. Explain the shape of this graph.   (4)

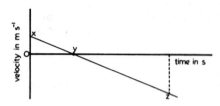

M6
The motions of three vehicles are displayed on the velocity-time
graph below. Each vehicle is known to be travelling in a straight
line.

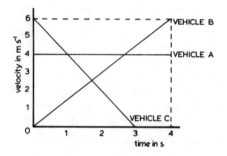

(a) Describe in general terms the motion of each vehicle.   (2)
(b) Calculate the acceleration of each vehicle.   (3)
(c) Which vehicle travels farthest during this section of their
     motions?   (3)
(d) What is the average velocity of each vehicle?   (2)

M7
The tape shown below is produced by a ticker-tape timer operating
at 50 Hz while a truck, to which the tape is attached, accelerates
down a sloping runway.

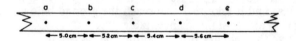

(a) Calculate the acceleration of the truck by comparing
     sections **ab** and **bc** of the tape.   (3)
(b) Calculate the acceleration of the truck by comparing
     sections **ab** and **de** of the tape.   (3)
(c) What is the average velocity of the truck during section
     ae of the tape?   (2)
(d) Construct a histogram to represent the motion of the truck. (2)

M8

A stroboscope photograph is taken of a ball as it drops on to a bench, using a flash-rate of 20 f.p.s. The distances shown on the diagram are actual distances, in cm, fallen by the ball in equal time intervals.

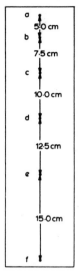

(a) Calculate a value for the acceleration due to gravity from the above information. (6)

(b) Estimate the velocity of the ball at position **f**, which is just above the surface of the bench. (4)

M9

John places a trolley on a friction-compensated track and attaches an elastic cord to it as shown.

(a) He then releases the trolley and moves it along the track, taking care to ensure that the cord is always stretched by the same amount. Describe the motion produced. (2)

(b) While the trolley is moving along, John suddenly lets go of the cord so that it slackens completely. Describe the subsequent motion of the trolley. (2)

(c) Draw a velocity-time graph for the trolley to represent its motion over sections (a) and (b). (2)

(d) Draw a graph to show how the force acting upon the trolley varies with time over sections (a) and (b) (2)

(e) John wishes to double the force which he applies to the trolley.
Describe how he could do this. (2)

M10

The following table indicates the forces applied to various trolleys.
What acceleration is produced in each of the arrangements (b) to
(f)?   (5 x 2)

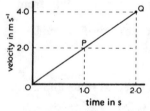

	TROLLEY MASS (units)	FORCE ARRANGEMENT (units)	ACCELERATION PRODUCED (units)
(a)	1	→F=1	6
(b)	1	→F=2	?
(c)	2	→F=1	?
(d)	2	→F=2	?
(e)	3	F=1← →F=2	?
(f)	3	F=2← →F=2	?

M11

The following velocity-time graph refers to the motion of a truck of
mass 0.80 kg which has a constant force applied to it on a friction-
compensated track.

(a)   Explain how the track can be compensated for friction.   (2)
(b)   Calculate the acceleration of the truck, in m s^{-2}.   (2)
(c)   What is the size of the force applied to it?   (2)
(d)   Compare the distances travelled by the truck during
        sections OP and PQ.   (2)
(e)   If the force is suddenly removed at t = 2.0 s, how far will
        the truck move in the next second?   (2)

M12

During a space training programme, an astronaut of mass 80 kg is
strapped into a rocket sledge of mass 220 kg which is mounted on a
long, frictionless rail.

The sledge is accelerated rapidly from rest by turning on the rocket motors for 2.0 s. It then moves at 100 m s^{-1} along the rail for the next 7.0 s, and finally is decelerated to rest by the application of a sudden thrust from the retro-motors for 4.0 s. Assuming that the mass of fuel ejected is negligible compared to the mass of the sledge,

(a) Calculate (i) the forward thrust, and (ii) the retro-thrust exerted on the sledge and passenger. (4)

(b) Estimate the total length of the journey along the rail (6)

M13

Mr. Fortune's car breaks down while he is driving his family through a safari park and so he persuades one of the local elephants to push his car.

Mass of car and passengers = 1200 kg

(a) The car moves forward with a constant velocity of 1.0 m s^{-1} when the elephant applies a constant horizontal force of 600 N to it. What must be the value of the friction force opposing the motion of the car? (2)

(b) Assuming that this friction force remains constant, what horizontal force would the elephant have to apply to make the car accelerate at 1.5 m s^{-2}? (4)

(c) While the elephant is applying a force of 2400 N, Miss Fortune jumps out of the car to clap the tigers and the acceleration of the car increases to 1.6 m s^{-2}. What was the mass of Miss Fortune? (4)

M14

A ball-bearing X is rolled down a ramp and projected horizontally over the edge of a bench. Just as it leaves the edge, another ball-

bearing Y is released from the same height by an electromagnet.
A stroboscope photograph is taken of the subsequent motion,
using a flash-rate of 10 f.p.s.
Four images of Y are included in the diagram and the distances
shown are the actual distances fallen by Y between flashes.

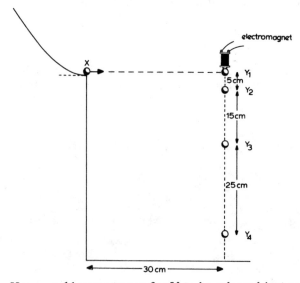

(a) How would you arrange for Y to be released just as X
     leaves the edge of the bench?   (3)
(b) The balls collide at position Y_4.
     (i)   Copy the above diagram and then mark in the
           position of the four images of ball-bearing X.   (3)
     (ii) Calculate the speed at which X left the edge of the
           bench.   (2)
(c) What would you expect to happen when X is moving
     faster as it leaves the bench?   (2)

M15

The members of a school rowing crew are asked to find out the mass
of their rowing boat. The boat is far too long (12.8 m) to be weighed
directly and so they borrow 4 weighing machines of the compression-
spring type, each with a range of 0–1000 N.

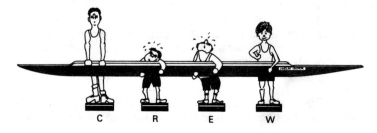

When the crew holds the boat as shown, the readings recorded on the balances are as detailed in the last column of this table.

Balance	Reading with oarsman only	Reading when holding boat
C	550 N	700 N
R	600 N	950 N
E	650 N	950 N
W	600 N	800 N

(a) What is the mass of the boat in kg? (4)
(b) Why do the others warn member W not to let go? (2)
(c) Describe what would happen to the readings on the balance if the crew, at a given signal from the cox, raised the boat 'above-heads'. Explain this effect in terms of Newton's third law. (4)

M16
Trolley A of mass 0.80 kg is travelling to the right along a level track at 6.0 m s^{-1} just before it collides with another trolley B of mass 1.20 kg. The trolleys become coupled together during the impact.

(a) How would you arrange for the trolleys to couple together upon impact? (2)
(b) What name is given to this kind of collision? (2)
(c) Calculate the common velocity of the trolleys after impact for each of the following cases:
   (i) trolley B is stationary before the impact
   (ii) trolley B is moving at 4.0 m s^{-1} **to the left** before impact,
   (iii) trolley B is moving at 4.0 m s^{-1} **to the right** before impact. (6)

M17
Sam Squintshott attempts to find out the speed at which a pellet leaves his trusty air-rifle by firing it horizontally into a block of plasticene (Y), which is fixed securely to a vehicle (X). The pellet embeds in Y, and X then travels horizontally along the friction-free linear air track. Before the shot is fired, X is at rest on the air-track.

Sam records the following experimental information:

mass of pellet   = 0.010 kg
mass of X   = 0.130 kg
mass of Y   = 0.260 kg
distance AB   = 1.20 m
time for vehicle to
move from A to B   = 0.60 s

(a)   What is the total momentum of the objects moving along the
      track?   (3)
(b)   What is the momentum of the pellet just before it enters the
      plasticene?   (2)
(c)   Calculate the muzzle-velocity of the pellet.   (2)
(d)   Explain briefly how Sam could time the vehicle over section
      AB.   (3)

M18

(a)   Trolley P is placed against a spring-loaded trolley Q on a level
      bench as shown. The mass of trolley P is 1.2 kg.

When the spring is released by pushing the plunger, P moves
to the left at 3.0 m s^{-1} and Q moves to the right at 2.0 m s^{-1}.
What is the mass of trolley Q?   (4)

(b)   The mass of Sam Squintshott's latest rifle is 8.00 kg. If it
      fires a 10 gram bullet at 400 m s^{-1}, estimate the initial recoil
      velocity of this rifle.   (3)
(c)   Explain briefly how a rocket works.   (3)

M19

Two eskimo boys, Ig and Lew, have been learning about momentum
at school and they decide to do some tests of their own outdoors.
They modify their sledges by putting large buffer-springs on the
front, and then they sit facing each other on a polished sheet of ice.
A stretched spring is connected between the sledges and they are
prevented from moving by the moorings at X and Y. Each modified
sledge has a mass of 20 kg.

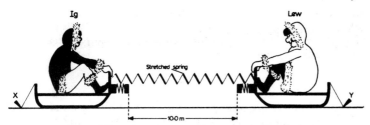

At a given signal, the moorings are released and the sledges move towards each other. Ig moves 6.0 metres and Lew moves 4.0 metres before the sledges collide.

(a) What is the total momentum of this system
    (i) just before the moorings are released
    (ii) just before the buffer-springs meet?   (2)

(b) If Ig has a mass of 40 kg, what is the mass of Lew?   (3)

(c) Assuming negligible friction, how does the force on Ig's sledge compare with that on Lew's
    (i) while the elastic cord is stretched
    (ii) while it is slack
    (iii) while the buffer-springs are compressed during the impact?   (3)

(d) Just before the impact, Ig is moving at 3.0 m s^{-1}. What is Lew's speed at this time?   (2)

M20

(a) Rory McBlister is instructed to carry an important package from his base-camp (X) to the summit of Ben Phew (Y) as quickly as possible. He travels a total distance of 2700 m up the mountainside and rises 1200 m above X in a record time of 40 minutes. It is known that Rory weighs 800 N and that the package weighs 200 N.

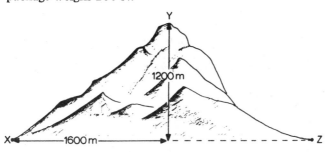

    (i) How much work does Rory do against gravity?   (2)
    (ii) What is the potential energy of the package at Y, if it is assumed to be zero at X?   (2)
    (iii) Calculate the average power developed by Rory during his climb.   (3)

(b) His friend, Jock McCrafty, is waiting at the summit to collect this important package and, when Rory arrives, he

thanks him for delivering his lunch so promptly! In a rage,
Rory chases Jock down the other side of the mountain to Z.
If Jock loses 9.0×10^5 J of potential energy during his hurried
descent, what is his mass?   (3)

M21

Mr. Fortune is going overseas for a well-deserved holiday and he stands
on the quayside watching a crane lift his car 9.8 metres up above
ground level. The crane takes 16 s to lift the car, whose mass is 800 kg.

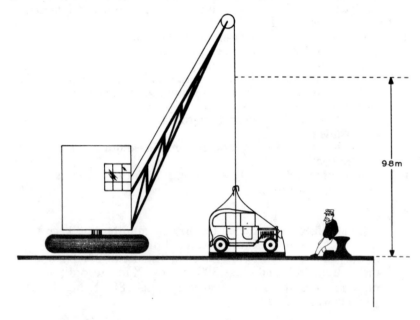

(a) What is the weight of his car?   (1)
(b) What is the tension in the cable while the car is hovering
    above the quayside?   (1)
(c) How much potential energy does the car gain during the
    lift?   (2)
(d) Calculate the useful power output of the crane.   (2)
(e) Explain why the total power input to the crane, which is
    electrically driven, is considerably greater than the value
    for (d).   (2)
(f) Unfortunately, the cable snaps and Mr. Fortune sees his car
    crash down into the ground. Estimate the maximum speed
    reached by the car during its fall.   (2)

M22

Anna 'free-wheels' down slope XY on her bicycle and then continues
to 'free-wheel' along the level road until she stops at Z.

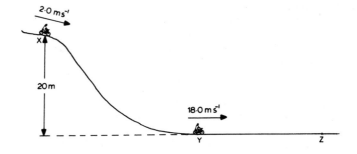

Anna's mass	= 40 kg	position	X	Y	Z
bicycle's mass	= 20 kg	speed (in m s⁻¹)	2.0	18.0	0

(a) How much potential energy is lost by Anna and her bicycle in travelling from X to Y? (2)

(b) How much kinetic energy is gained by Anna and her bicycle in travelling from X to Y? (2)

(c) Account for the difference between answers (a) and (b). (3)

(d) If the distance YZ is 30 metres, calculate the average decelerating force acting during this section. (3)

M23

Here is a list of quantities commonly used during the study of mechanics: momentum, time, distance, acceleration, energy, volume, velocity, weight, mass, force, speed, displacement. The following table has a column for scalar quantities and a column for vector quantities. Make up two lists, each containing 6 quantities, one for scalars and the other for vectors. The quantities should be listed alphabetically. The first row of the table has already been filled in to help you.

Scalar	Vector
distance	acceleration

(10)

Solutions to Worked Examples

M1

(a) The disc would appear stationary with a single line because the line rotates twice between flashes.

(b) The line would appear to be moving slowly forwards (clockwise) because the line rotates just over one revolution between flashes.

(c) The disc would appear stationary with a single line because the disc rotates once between flashes.

(d) The line would appear to be moving slowly backwards (anti-clockwise) because the line rotates by just less than one revolution between flashes.

(e) The disc would appear stationary with two lines because the line only moves round by one half of a revolution between flashes.

M2

(a) Each **space** between dots represents a time interval of $\frac{1}{50}$ th of a second (0.02 s). Since there are 5 such **spaces** between X and Y, the required time interval is 0.10 s. ($5 \times \frac{1}{50}$). Notice that we always count the **spaces** rather than the dots in this type of problem. Obviously,

number of dots = number of spaces + 1

(b)
$$\text{number of spaces} = \frac{\text{total time of experiment}}{\text{time to produce one space}} = \frac{5.0 \text{ s}}{0.02 \text{ s}} =$$

$$\frac{500}{2} = 250$$

number of dots on tape = 250 + 1 = 251

(c) There are 6 **spaces** in section OP and so the time for one space is $\dfrac{0.30 \text{ s},}{6}$ or 0.05 s.

$$\text{since frequency} = \frac{1}{(\text{time between dots})} = \frac{1}{0.05} = \frac{100}{5} = 20, \text{ then}$$

the frequency of this timer is 20 Hz.

M3

(a) Rory's displacement is represented by a vector drawn straight from X to Y.

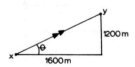

If this diagram is drawn accurately to scale, $\overline{XY}$ represents 2000 m at an angle $\theta = 37°$.

(b) $\quad$ average speed $= \dfrac{\text{total distance}}{\text{time taken}} = \dfrac{2400 \text{ m}}{(100 \times 60) \text{ s}} = 0.4 \text{ m s}^{-1}$

(c) $\quad$ average velocity $= \dfrac{\text{total displacement}}{\text{time taken}} = \dfrac{2000 \text{ m}}{(100 \times 60) \text{ s}}$

$$= \frac{1}{3} \text{ m s}^{-1}, \text{ at an angle } \theta = 37°$$

(d) Jock climbs for 1 hour (1 hour 40 min − 40 min) at 0.5 m s^{-1}
distance travelled = average speed x time = 0.5 x 60 x 60 =
1800 m.

M4

(a) The car is increasing its velocity by 4 m s^{-1} during each second
∴ its acceleration is 4 m s^{-2}, or use

$$a = \frac{(v - u)}{t} = \frac{(16 - 0) \text{ m s}^{-1}}{4 \text{ s}} = 4 \text{ m s}^{-2}$$

(b) again, use

$$a = \frac{(v - u)}{t} = \frac{(0 - 2.4) \text{ m s}^{-1}}{0.8 \text{ s}} = -3.0 \text{ m s}^{-2} \text{ (The minus sign}$$

means deceleration.)

(c) (i) $\quad v = u + at = 0 + (10 \times 0.4) = 4 \text{ m s}^{-1}$ $\qquad$ downwards
(ii) $\quad v = u + at = 8 + (10 \times 0.4) = 12 \text{ m s}^{-1}$ $\qquad$ downwards
(iii) $\quad v = u + at = 8 - (10 \times 0.4) = 4 \text{ m s}^{-1}$ $\qquad$ upwards

M5

(a) (i) Since

$$\text{acceleration} = \frac{\text{change in velocity}}{\text{time taken}}, \quad a = \frac{(v - u)}{t}$$

∴ $\quad v - u = at, \quad$ or $\quad v = at, \quad$ since $\quad u = 0 \text{ m s}^{-1}$

when pebble is dropped from rest. Thus $v = 10 \times 3.2 = 32 \text{ m s}^{-1}$.
(ii) The pebble accelerates uniformly from 0 m s^{-1} to 32 m s^{-1}
∴ its average velocity during the drop is 16 m s^{-1}
Now displacement = average velocity x time
∴ height dropped $\quad = 16 \times 3.2 = 51.2$ metres.

(b) Time would be the same, irrespective of mass, provided air
resistance is negligible.

(c) At X, the pebble has its maximum upwards (positive) velocity.
Between X and Y, the pebble is decelerated uniformly to rest.
At Y, the object is at rest for an instant at its highest point.
Between Y and Z, the pebble is accelerated uniformly down-
wards (negative direction). At Z, the pebble is just about to
strike the ground.

M6

(a)　Vehicle A is travelling at a constant velocity.
Vehicle B is travelling with a uniform acceleration.
Vehicle C is travelling with a uniform deceleration.

(b)　$\text{acceleration} = \dfrac{\text{change in velocity}}{\text{time taken}}$

$\qquad\qquad = \dfrac{(\text{final velocity} - \text{initial velocity})}{\text{time taken}}$

or simply

$$a = \frac{(v - u)}{t}$$

For vehicle A

$$a = \frac{(v - u)}{t} = \frac{(4 - 4)\text{m s}^{-1}}{4\text{ s}} = 0 \text{ m s}^{-2} \text{ (no acceleration)}$$

For vehicle B

$$a = \frac{(6 - 0)\text{ m s}^{-1}}{4\text{ s}} = 1.5 \text{ m s}^{-2}$$

For vehicle C

$$a = \frac{(0 - 6)\text{ m s}^{-1}}{3\text{ s}} = -2 \text{ m s}^{-2} \text{ (minus sign indicates deceleration)}$$

(c)　distance gone in straight line = displacement = area under v/t graph.
For vehicle A: displacement, $s_A = (4 \times 4) = 16$ m
For vehicle B: displacement, $s_B = (\frac{1}{2} \times 4 \times 6) = 12$ m
For vehicle C: displacement, $s_C = (\frac{1}{2} \times 3 \times 6) = 9$ m
∴ vehicle A travels farthest.

(d)　$\text{average velocity} = \dfrac{\text{displacement}}{\text{time}}, \quad \text{or} \quad v = \dfrac{s}{t}$

Vehicle A:

$$v_A = \frac{16 \text{ m}}{4 \text{ s}} = 4 \text{ m s}^{-1} \text{ (as expected)}$$

Vehicle B:

$$v_B = \frac{12 \text{ m}}{4 \text{ s}} = 3 \text{ m s}^{-1}$$

Vehicle C:

$$v_C = \frac{9 \text{ m}}{3 \text{ s}} = 3 \text{ m s}^{-1}$$

M7

(a)　Average velocity of truck during time interval **ab** $= \dfrac{5.0 \text{ cm}}{0.02 \text{ s}}$

$\qquad\qquad = 250 \text{ cm s}^{-1}$

The truck is actually travelling at 250 cm s^{-1} at the middle of the time interval **ab** i.e. 0.01 s after dot **a** is produced and 0.01 s before dot **b** is produced.

Average velocity of truck during time interval **bc** $= \dfrac{5.2 \text{ cm}}{0.02 \text{ s}}$
= 260 cm s^{-1}.

The truck is actually travelling at 260 cm s^{-1} at the middle of the time interval **bc** i.e. 0.01 s after dot **b** is produced and 0.01 s before dot **c** is produced.

Change in velocity of truck = (260 − 250) = 10 cm s^{-1}

Time for this change = 0.02 s (from the middle of time interval **ab** to the middle of time interval **bc**).

$\therefore$ acceleration $= \dfrac{\text{change in velocity}}{\text{time for change}} = \dfrac{10 \text{ cm s}^{-1}}{0.02 \text{ s}} = 500$ cm s^{-2}

truck is accelerating at 5.00 m s^{-2}

(b) average velocity during **ab** $= \dfrac{5.0}{0.02} = 250$ cm s^{-1}

average velocity during **de** $= \dfrac{5.6}{0.02} = 280$ cm s^{-1}

acceleration $= \dfrac{(280 - 250)}{3 \times 0.02} = \dfrac{30}{0.06} = 500$ cm s^{-2}, or 5.00 m s^{-2}

(c) Average velocity during **ae** $= \dfrac{(5.0 + 5.2 + 5.4 + 5.6) \text{ cm}}{4 \times 0.02 \text{ s}}$

$= 265$ cm s^{-1}

or 2.65 m s^{-1}.

(d) Cut tape into sections **ab, bc, cd** and **de** and then mount the sections side by side. The result is a histogram showing the features of a uniformly accelerating motion.

M8

(a) The time interval between adjacent images $= \frac{1}{20}$ s = 0.05 s. Since the distance between images is increasing by 2.5 cm each time, the acceleration is **uniform**, and so we need only consider the first two time intervals.

average velocity during **ab** $= \dfrac{5.0}{0.05} = 100$ cm s^{-1}

average velocity during **bc** $= \dfrac{7.5}{0.05} = 150$ cm s^{-1}

change in velocity = $(150 - 100) = 50$ cm s^{-1}
time for change = 0.05 s

$$\text{acceleration} = \frac{\text{change in velocity}}{\text{time}} = \frac{50 \text{ cm s}^{-1}}{0.05 \text{ s}} = 1000 \text{ cm s}^{-2}$$

∴ acceleration due to gravity = 10.0 m s^{-2}

(b)   Average velocity during **ef** $= \dfrac{15.0 \text{ cm}}{0.05 \text{ s}} = 300$ cm s^{-1}.

The ball is travelling at 300 cm s^{-1} at the middle of the time interval **ef**. We want the velocity at the end of this interval i.e. $\frac{1}{40}$ s later. Since the acceleration is 1000 cm s^{-2}, it gains 25 cm s^{-1} during this time interval ∴ required velocity = 300 cm s^{-1} + 25 cm s^{-1} = 325 cm s^{-1}, or 3.25 m s^{-1}.

M9

(a)   The trolley would accelerate uniformly because it is acted upon by a constant, unbalanced force due to the elastic cord.

(b)   The trolley would then move along the track with a constant velocity because there is now no unbalanced force acting upon it. **Note.** If the track had not been friction-compensated, the friction force on the trolley would have decelerated it to rest.

(c)

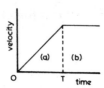

(d)

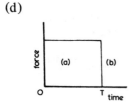

(e)   Another identical elastic cord should be placed 'in parallel' with the first one and both should be stretched by the same amount as before.

M10

The important points to remember are
    (i)   the acceleration is directly proportional to the resultant applied force.

(ii) the acceleration is inversely proportional to the mass of the trolley. i.e.

$$a \propto \frac{F}{m}, \quad \text{or} \quad a = k \times \frac{F}{m} \quad (k \text{ is a constant})$$

Now, from (a) we know that

$$6 = k \times \frac{1}{1} \quad \therefore k = 6 \text{ in this problem.}$$

using $a = 6 \times \dfrac{F}{m}$, we obtain

(b) $a = 6 \times \dfrac{2}{1} = 12$ units (double the force, double the acceleration)

(c) $a = 6 \times \dfrac{1}{2} = 3$ units (double the mass, half the acceleration)

(d) $a = 6 \times \dfrac{2}{2} = 6$ units (double both mass and force, no change in acceleration)

(e) $a = 6 \times \dfrac{(2-1)}{3} = 2$ units (since resultant force is only 1 unit)

(f) $a = 6 \times \dfrac{(2-2)}{3} = 0$ units (since there is zero resultant force)

M11

(a) The truck is given a short, sharp push on a slightly tilted track.
If the truck slows down, the tilt must be increased.
If the truck speeds up, the tilt must be decreased.
The angle of tilt is correct if the truck continues at constant velocity after the initial short, sharp push.

(b) $\qquad \text{acceleration} = \dfrac{\text{change in velocity}}{\text{time taken}} = \dfrac{(v - u)}{t}$

$$= \frac{4.0 \text{ m s}^{-1}}{2.0 \text{ s}} = 2.0 \text{ m s}^{-2}$$

(c) By Newton's second law, $F = ma = 0.80 \times 2.0 = 1.6$ N

(d) Displacement = Area under v/t graph.
Displacement during OP = ($\frac{1}{2}$ x 1.0 x 2.0) = 1.0 m.
Displacement during PQ = (1.0 x 2.0) + ($\frac{1}{2}$ x 1.0 x 2.0) = 3.0 m.

(e) The truck will move at a constant velocity of 4.0 m s^{-1} when the force is removed $\therefore$ it travels 4.0 m in the next second.
In the absence of resultant external forces, an object does not alter in velocity. (This is one way of saying Newton's first law.)

M12

(a) (i) $a = \dfrac{(v - u)}{t} = \dfrac{(100 - 0)}{2.0} = 50 \text{ m s}^{-2}$

forward thrust, $F = ma = (220 + 80) \times 50 = 15\,000$ N, or 1.5×10^4 N

(ii) $a = \dfrac{(v - u)}{t} = \dfrac{(0 - 100)}{4.0} = -25 \text{ m s}^{-2}$ (minus means deceleration)

retro-thrust, $F = ma = (220 + 80) \times (-25) = -7500$ N, or $F = -7.5 \times 10^3$ N.

(b) Part I: Rocket increases the velocity of the sledge from 0 m s^{-1} to 100 m s^{-1}. If we assume that acceleration is uniform, then average velocity is 50 m s^{-1}.
displacement = average velocity x time = 50 x 2.0 = 100 m
Part II: Rocket travels at 100 m s^{-1} for 7.0 s.
displacement = 100 x 7.0 = 700 m
Part III:
displacement = 50 x 4.0 = 200 m
total length of journey = (100 + 700 + 200) = 1000 m.

M13

(a) Since the car is neither accelerating nor decelerating, the resultant horizontal force on it must be zero. Thus, the friction force must be 600 N acting to the right to cancel the effect of the applied force.

(b) Let the force applied by the elephant be X N and let the friction force be Y N. Then from Newton's second law, $F = ma$
$(X - Y) = ma = 1200 \times 1.5 = 1800$ N
∴ $(X - 600 \text{ N}) = 1800$ N
∴ $X = 2400$ N acting to left

(c) Let new total mass of car and remaining passengers be M kg, then $(X - Y) = Ma$
∴ $(2400 - 600) = M \times 1.6$

$$M = \frac{1800}{1.6} = 1125 \text{ kg}$$

mass of Miss Fortune = original mass − final mass = (1200 − 1125) = 75 kg

M14

(a) Place a small gate made of aluminium foil across the ramp at the edge of the bench so that X will break open the gate. If the gate is used as an electrical switch and wired up in series with the electromagnet and a suitable power supply, the power to the electromagnet will be cut off automatically at the required time.

(b) (i) X_2 should be 10 cm out from bench and at same level as Y_2
X_3 should be 20 cm out from bench and at same level as Y_3
X_4 should be at the same position as Y_4 (collision point)

(ii) X moves sideways by 30 cm during the production of four flashes i.e. during three flash-intervals. Each flash interval is 0.1 s.

∴ X moves 30 cm sideways in 0.3 s

∴ horizontal velocity $= \dfrac{30 \text{ cm}}{0.3 \text{ s}} = 100 \text{ cm s}^{-1} = 1.00 \text{ m s}^{-1}$

(c) Ball X would still hit ball Y, but the collision would occur earlier i.e. nearer electromagnet.

M15

(a) total weight of crew $= (550 + 600 + 650 + 600)$ N $= 2400$ N
total weight of crew and boat $= (700 + 950 + 950 + 800)$ N
$= 3400$ N
weight of boat alone $= (3400 - 2400) = 1000$ N

but weight $= mg$ ∴ $m = \dfrac{\text{weight}}{g} = \dfrac{1000}{10} = 100$ kg

Mass of rowing boat is 100 kg.

(b) If W lets go, the others must share an additional downward force of $(800 - 600)$ N, or 200 N. Assuming that this is distributed equally, the forces on balances R and E would exceed the maximum range (1000 N).

(c) While the boat is accelerating upwards, the balance readings would increase, but revert to their previous values whenever the boat was held still again. Newton's third law states that 'to every action, there is an equal and opposite reaction'. In this case the action is the additional upwards force exerted by crew on boat and the reaction is the additional downwards force by boat on crew.

M16

(a) Attach a pin to the front of trolley A and fix a piece of cork to B as shown so that they couple together on impact.

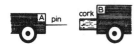

(b) An inelastic collision, since some kinetic energy is transformed into heat, sound, etc. during impact.

(c) Let both trolleys move at x m s^{-1} after impact.

(i) total momentum of system $=$ total momentum of system
just before impact   just after impact

$$m_A u_A + m_B u_B = (m_A + m_B) x$$
$$(0.80 \times 6.0) + (1.20 \times 0) = (0.80 + 1.20) x$$
$$4.8 + 0 = 2.0 x$$

this leads to $x = \dfrac{4.8}{2.0} = 2.4$ m s^{-1} to right

(ii) As above, total momentum before = total momentum after

$$m_A u_A - m_B u_B = (m_A + m_B) x$$

(The minus sign indicates that B is moving to left.)

$$4.8 - (1.20 \times 4.0) = 2.0 \, x$$
$$4.8 - \qquad 4.8 \quad = 2.0 \, x$$
$$\therefore \quad x = 0 \text{ m s}^{-1}$$

Thus the trolleys are brought to rest by the collision. This only happens when they have equal and opposite momenta before impact.

(iii) again, total momentum before = total momentum after

$$m_A u_A + m_B u_B = (m_A + m_B) x$$
$$4.8 + (1.20 \times 4) = 2.0 \, x$$

leading to $x = 4.8$ m s^{-1} to right

M17

(a) total momentum = total mass × velocity

$$= \text{total mass} \times \frac{\text{distance AB}}{\text{time}}$$

$$= (0.010 + 0.130 + 0.260) \text{ kg} \times \frac{1.20 \text{ m}}{0.60 \text{ s}}$$

$$= 0.80 \text{ kg m s}^{-1}$$

(b) By principle of conservation of momentum, pellet's momentum is also 0.80 kg m s^{-1}

mass of pellet × muzzle velocity = 0.80 kg m s^{-1}

$$0.010 \times V = 0.80 \Rightarrow V = \frac{0.80}{0.010} = 80 \text{ m s}^{-1}$$

(d) Use photocells and light beams at A and B, arranged so that the leading edge of X starts an accurate timer as it passes A and stops it when it passes B, **or**, use stroboscope photograph method.

M18

(a) total momentum before release = 0 kg m s^{-1}

∴ total momentum after release = 0 kg m s^{-1}

taking motion to right as positive and to the left as negative

$$(m_Q \times v_Q) - (m_P \times v_P) \qquad = 0$$
$$(m_Q \times 2.0) - (1.2 \times 3.0) = 0$$
$$\therefore \quad m_Q = \frac{3.6}{2.0} = 1.8 \text{ kg}$$

trolley Q has a mass of 1.8 kg

(b) as in (a), giving

$$8.00 \times v_R = 0.010 \times 400$$

$$\therefore \quad v_R = \frac{4.00}{8.00} = 0.50 \text{ m s}^{-1}$$

rifle recoils at 0.50 m s^{-1}

(c) Just before the rocket is fired, its momentum is zero. When fired hot gases are ejected from the rocket. The rocket will have to move in the opposite direction so that momentum is conserved.

M19

(a) (i) Since both sledges are at rest at start, the total momentum of system is zero.

(ii) Since total momentum is conserved, the momenta must be equal and opposite at every stage of their approach to make total momentum zero.

(b) Let X kg be mass of Lew. From (a) (ii), we see that
(momentum of Ig and sledge) — (momentum of Lew and sledge)

$$= 0$$
$$(40 + 20) \times 6.0 - (X + 20) \times 4.0 = 0$$
$$360 - 4.0 (X + 20) = 0$$
$$\therefore \quad X = 70 \text{ kg.}$$

(c) By Newton's third law, the forces on the sledges are equal and opposite in cases (i) and (iii). In case (ii), there is no force as the cord is slack.

(d) Let Lew's speed be V m s^{-1}, then from (b), we see

$$(40 + 20) \times 3.0 - (70 + 20) \times V = 0$$

$$V = \frac{60 \times 3.0}{90} = 2.0 \text{ m s}^{-1}.$$

M20

(a) (i) work done = force x distance moved in direction of force
= total weight x height raised
= 1000 N x 1200 m
= 1.2×10^6 J

(ii) potential energy of package = mgh = 200 N x 1200 m
= 2.4×10^5 J

(iii) average power = $\dfrac{\text{total work done}}{\text{time taken}} = \dfrac{1.2 \times 10^6 \text{ J}}{40 \times 60 \text{ s}}$
= 500 J s^{-1} = 500 W.

(b) loss in P.E. = mgh = 9.0×10^5 J

$$\therefore \quad m \times 10 \times 1200 = 9.0 \times 10^5$$

$$\therefore \quad m = \frac{9.0 \times 10^5}{1.2 \times 10^4} = 75 \text{ kg}$$

Jock has a mass of 75 kg and a weight of 750 N (weight = mg)

M21

(a)   weight = force of gravity on car = mg = 800 x 10 = 8000 N

(b)   Tension in cable must be equal in size and opposite in direction to the weight of the car for equilibrium

$\therefore$ tension is 8.0×10^3 N.

(c)   increase in P.E. = mgh = 800 x 10 x 9.8 = 78 400 = 7.84×10^4 J

(d)   useful power output = $\dfrac{\text{potential energy}}{\text{time}} = \dfrac{78\ 400 \text{ J}}{16 \text{ s}}$

= 4900 W (4.9 kW)

(e)   Additional energy has to be supplied to raise the platform and the cable and to overcome the forces of friction in the machinery. Heat energy will also be produced in the electrical parts of the crane.

(f)   final K.E. of car = initial P.E. of car

$$\tfrac{1}{2}\, m\, v^2 = mgh$$
$$v^2 = 2gh$$
$$v = \sqrt{2gh} = \sqrt{2 \times 10 \times 9.8} = \sqrt{196} = 14 \text{ m s}^{-1}$$

M22

(a)   loss in P.E. = mgh = (40 + 20) x 10 x 20 = 12 000 J
(or 1.2×10^4 J)

(b)   gain in K.E. = K.E. at Y $-$ K.E. at X
$$= \tfrac{1}{2} \times 60 \times (18.0)^2 - \tfrac{1}{2} \times 60 \times (2.0)^2$$
$$= 30(324 - 4)$$
$$= 9600 \text{ J (or } 9.6 \times 10^3 \text{ J)}$$

(c)   The difference between (a) and (b) is (12 000 $-$ 9600) J, or 2400 J. This energy has been transformed into other types of energy e.g. energy needed to rotate the wheels, heat generated by friction forces between the moving parts of the bicycle, between the tyres and the ground etc. Some sound is created from the noisy bike parts.

(d)   work done against friction force = loss in K.E. along YZ
friction force x distance $= \tfrac{1}{2} \times 60 \times (18.0)^2$
$F \times 30 = 30 \times 324$ $\therefore$ $F = 324$ N

M23

Scalar	Vector
energy	displacement
mass	force
speed	momentum
time	velocity
volume	weight

Practice Questions

1. (M1)
A circular disc has a single line painted along its radius. It is attached to the axle of an electric motor. When the motor is switched on, the disc is rotated at 12 revolutions per second.
Describe the appearance of the disc under stroboscope lighting at each of the following flash rates:
(a) 12 f.p.s. (flashes per second)
(b) 24 f.p.s.
(c) 36 f.p.s.
(d) 6 f.p.s.
(e) 3 f.p.s.

2. (M2)
A ticker-tape is moved through a ticker-timer for 8.0 s as measured on a stopwatch. If the ticker-timer is operating at 50 Hz, how many dots would you expect to be printed on the tape?

3. (M2)
When a ticker-timer is operated from a signal generator, it produces 401 dots in 10.0 s. Calculate
(a) the time interval between neighbouring dots,
(b) the frequency setting of the signal generator.

4. (M3)
Slumbertown lies 15 km from its sister-town Yawnville, across a flat desert. The road joining the towns goes east from Yawnville for 12 km and then goes north until it reaches Slumbertown.
(a) What is the length of the northwards section of the road?
(b) The fastest long-distance runner in the district runs from Yawnville to Slumbertown by road in 2 hours 55 minutes.
   (i) Calculate his average speed in m s^{-1}.
   (ii) What is his total displacement?
   (iii) What is his average velocity?

5. (M4)
The table below gives the velocity of a racing car at seconds intervals as it accelerates from rest along a straight track.

time (in s)	0	1	2	3	4
velocity (in m s^{-1})	0	8	16	24	32

(a) Calculate the acceleration of the racing car.
(b) What is its average velocity?
(c) What is its velocity 1.5 s after the start?

6. (M4)
A girl on a skateboard travels along a level track and then moves up

a slope until she has decelerated to rest. Her speed is recorded at intervals of 0.5 s. The results are as follows

time (in s)	0	0.5	1.0	1.5	2.0	2.5	3.0	3.5	4.0
speed (in m s^{-1})	5	5	5	5	4	3	2	1	0

(a) At what speed did she travel along the level track?
(b) At what rate did she decelerate on the slope?

7. (M5)
An object, dropped from the roof of a skyscraper, takes 9.0 s to reach the ground. Using the acceleration due to gravity as 10 m s^{-2}, and neglecting air resistance, estimate
(a) the velocity of the object just before it hits the ground,
(b) the height of the skyscraper.

8. (M6)
Three velocity-time graphs are drawn to represent the motions of three objects. Object A is travelling with a constant velocity. Object B is travelling with a constant acceleration. Object C is travelling with a constant deceleration. Describe how the gradients of the three graphs would compare.

9. (M6)
(a) What does the area under a velocity-time graph represent?
(b) What does the gradient of a velocity-time graph represent?

10. (M6)
A vehicle on a straight road accelerates from rest at 2.0 m s^{-2} for 5 s. It then travels at a constant speed for the next 5 s. Finally, it decelerates uniformly to rest in a further 5 s.
(a) What is the top speed reached by the vehicle?
(b) Draw a velocity-time graph of the motion and use it to find the total distance travelled.
(c) What is the average speed
    (i) during the acceleration?
    (ii) during the deceleration?
    (iii) during the entire journey?

11. (M7)
As a trolley accelerates down a ramp, its motion is recorded on a ticker-tape which is attached to the trolley. The timer used operates at 50 Hz. The spacing between dots on a sample section of the tape is 1.5 cm, 1.7 cm, 1.9 cm, 2.1 cm and 2.3 cm. Calculate the acceleration of the trolley, in m s^{-2}.

12. (M8)
(a) A small ball-bearing is released from an electromagnet and a stroboscope photograph is taken as it falls freely. Draw a

sketch to indicate the type of photograph which you would
expect to be obtained.
(b)   The experiment is now repeated using a larger ball-bearing
but the same flash-rate. How would this photograph compare
with the one discussed in part (a)?

13. (M9)
During an experiment with a trolley on a runway, the teacher states
that it will be necessary to compensate for friction by tilting the
runway. Explain how this is done.

14. (M9)
A trolley is placed on a track which has been friction-compensated.
Describe the motion of the trolley after it has been given an initial
sharp push and then released. How would the motion differ if a
horizontal (uncompensated) track had been used instead?
Sketch velocity-time graphs for each type of motion.

15. (M10)
The following chart indicates the forces applied to various trolleys.
What accelerations are produced in arrangements (b) to (f)?

	(a)	(b)	(c)	(d)	(e)	(f)
trolley mass (units)	6	6	6	3	2	1
resultant force (units)	6	12	18	6	6	6
acceleration produced (units)	1	?	?	?	?	?

16. (M11)
A trolley moves along a level table against a constant frictional
force of 1 unit. What can you say about the applied horizontal force
if:
(a)   the trolley moves with a constant velocity?
(b)   the trolley moves with a constant acceleration?
(c)   the trolley moves with a constant deceleration?

17. (M11)
(a)   Newton's second law is often written in the form $F = ma$.
What units would you recommend for F, m and a?
(b)   What is the acceleration of an object of mass 3.0 kg when a
resultant force of 12.0 N acts upon it?
(c)   A trolley accelerates at 1.5 m s^{-2} when a force of 6.0 N is
applied to it. What must be the mass of this trolley?
(d)   A man of mass 80 kg is accelerating at 10 m s^{-2} as he dives
from a cliff edge into the sea. What force is the earth exerting
upon him? What name is given to this force due to gravity?

18. (M12)

During a space training programme, an astronaut of mass 90 kg sits in a rocket sledge of mass 410 kg and they move along a frictionless horizontal rail. The sledge is accelerated uniformly from rest for 3 s and reaches a top speed of 60 m s^{-1}. The motors are switched off for the next 4 s. Then the sledge is decelerated for 5 s and comes uniformly to rest.

(a)   What force accelerates the sledge and passenger during the first 3 seconds?

(b)   What force decelerates the sledge and passenger during the last 5 seconds?

(c)   What total distance does the sledge move along the rail?

19. (M13)

A motor car of mass 800 kg is travelling along a straight road at a constant speed of 20 m s^{-1}. The forward force of the engine is 1000 N.

(a)   What is the size of the friction force opposing the car's motion?

(b)   If the friction force stays the same and the forward force is increased to 1200 N, what acceleration does the car now have?

(c)   If the friction force stays the same and the forward force is decreased until the car decelerates at 0.5 m s^{-2}, what is the new value of the forward force?

20. (M14)

Alf and Fred are standing at the top of a cliff. Alf drops a rock and it takes 2.0 s to hit the beach below. Fred throws a stone horizontally outwards from the top of the cliff at 8.0 m s^{-1} and times it as it drops down to the sea.

(a)   What time does Fred's stone take to plunge down into the sea?

(b)   How far out from the base of the cliff is Fred's stone when it splashes down into the sea?

21. (M15)

(a)   Assuming that the acceleration due to gravity on Earth is 10 m s^{-2}, what is the weight of each of the following?

    (i)   a wrestler of mass 180 kg

    (ii)   a coin of mass 20 g

    (iii) a car of mass 1200 kg

(b)   On planet X, the acceleration due to gravity is only 5 m s^{-2}. What would be the weights of the wrestler. coin and car on planet X?

(c)   How would their masses compare on Earth and on planet X?

22. (M15)

State Newton's third law of motion and illustrate it in the following situations:

(a)   a footballer taking a penalty kick

(b)   a rocket during launching

23. (M16)

A truck of mass 1200 kg is travelling at 5.0 m s^{-1} when it collides with a stationary truck of mass 800 kg. The trucks move off, locked together, after collision.

(a) What is the total momentum of the system just before the impact?

(b) What is the total momentum of the system just after the impact?

(c) What is the common velocity of the trucks just after the impact?

(d) What name is given to this type of collision?

24. (M16)

Two wagons of equal mass are travelling along the same track in opposite directions and they crash 'head-on'. Wagon A was travelling to the right at 6.0 m s^{-1}, and wagon B was travelling to the left at 4.0 m s^{-1} before the collision.

(a) If the wagons couple together upon impact (inelastic collision), what is the velocity of the wagons after the impact?

(b) How much kinetic energy is converted into other forms of energy during the impact, if the mass of each wagon is 200 kg?

25. (M17)

A bullet of mass 50 g is moving at 100 m s^{-1} when it enters a stationary block of wood whose mass is 0.950 kg. The bullet becomes embedded in the block. Estimate the speed imparted to the block by the bullet, assuming that the entire motion is horizontal.

26. (M18)

Trolley A is a spring-loaded trolley of mass 1.6 kg. It is placed on a frictionless, level bench against trolley B whose mass is not known. When the spring is suddenly released, A moves to the left at 5.0 m s^{-1} and B moves to the right at 2.0 m s^{-1}.

(a) What is the total momentum of the system of two trolleys just after the spring is released?

(b) Calculate the mass of trolley B.

(c) Calculate the amount of kinetic energy produced by this 'explosion'.

27. (M18)

The mass of Colonel Niceshott's rifle is 5.00 kg and it fires a 20 gram bullet at 120 m s^{-1}. At what speed does the rifle recoil when the bullet is fired?

28. (M19)

Two pupils stand on skateboards holding the ends of a long spring. Friends pull the pupils outwards until the spring is stretched. At a given signal, the two pupils are released and they travel towards each other on their skateboards.

While the pupils are moving towards each other

(a)   how do the forces on them compare?
(b)   how do the momenta of the pupils compare?
(c)   how do their velocities compare?
(d)   what is the total momentum of the system?

29. (M20)

(a)   A horizontal force of 200 N is applied to a crate of mass 40 kg and it moves forward by 6 m.

    (i)   How much work is done by this force?

    (ii)   If the frictional force opposing the motion is 150 N, how much heat energy is produced as the crate is shifted?

(b)   A crane lifts a slab of concrete of mass 30 kg through a height of 20 m. How much potential energy does it give to the slab?

30. (M20)

During a training session, an athlete runs up a flight of stairs, through a vertical height of 40 m. If his mass is 80 kg and he takes 16 s, what average power does he develop?

31. (M21)

A hoist of mass 360 kg is loaded with 10 boxes each of mass 14 kg. The hoist is then raised at a constant vertical speed of 0.5 m s^{-1}.

(a)   What is the minimum power output of the motors which operate the hoist?

(b)   What is the tension in the hoist cable, in newtons?

UNIT H

Heat and the gas laws

Content List of Topic Areas

Worked Examples

H1

Jules does a 'warm-up' experiment by rotating a metal cylinder against a friction pad or belt as shown. He settles down to a steady turning rate of 40 revolutions per minute.

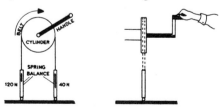

(a) The balance readings are 120 N and 40 N as shown and the diameter of the metal cylinder is 0.08 m.

(i)   What is the value of the frictional force acting round
the rim of the cylinder?   (2)

(ii)   What is the mechanical work done against the friction
belt in one minute?   (2)

(b)   What would be the effect upon the cylinder of this mechanical
work?   (1)

(c)   Jules went on to repeat the experiment with three different
energies supplied. The resulting temperature increases of the
cylinder are shown in the table:

energy supplied (E) (in joules)	2000	4000	6000
temperature rise (ΔT) (in $^\circ$C)	6	12	18

(i)   Suggest a method for measuring the temperature rise
ΔT of the metal.   (2)

(ii)   How is the temperature rise related to the energy
supplied?   (1)

(iii) Would you expect to get the same results if you changed
the mass of the metal, or the type of metal used for the
cylinder?   (2)

H2

Using the same form of apparatus as in the last question, Jules
adapted his experiment in two ways.

(a)   When he adapts it so that the rises in temperature of different
masses of the same metal can be compared, he obtains the
following results:

energy supplied (in J)	2000	2000	2000
mass of metal (in kg)	1.0	2.0	3.0
temperature rise (in $^\circ$C)	6.0	3.0	2.0

(i)   What two factors are kept constant this time?   (2)

(ii)   What is the relationship between the temperature rise
and the mass of the substance used?   (2)

(b)   When he adapts it so that the rise in temperature of different
metals can be compared, he obtains the following results:

energy supplied (in J)	2000	2000	2000
metal	X	Y	Z
mass of metal (in kg)	1.0	1.0	1.0
temperature rise (in $^\circ$C)	3.0	6.0	10.0

(i)   Which of the metals X, Y or Z is the one which was
used in part (a) of this question?   (1)

(ii)   Specific heat capacity is a measure of the amount of
energy needed to raise the temperature of 1 kg of a

substance by 1°C (or 1 K). Which metal has the
highest value of specific heat capacity, and which metal
has the lowest value? (2)

(iii) Calculate the specific heat capacity of metal Y. (3)

H3

(a) When a cylinder of aluminium of mass 5.0 kg receives 110 000 J
of energy, its temperature rises from 18°C to 43°C. Calculate
the specific heat capacity of aluminium, assuming that no energy
is lost to the surroundings. (3)

(b) The specific heat capacity of copper is 385 J $kg^{-1} \, °C^{-1}$ (or
J $kg^{-1} \, K^{-1}$) and its melting point is 1080°C. How much
energy would be needed to raise the temperature of a 20 kg
block of copper from 80°C up to its melting point? (4)

(c) Calculate the rise in temperature produced in an iron
cylinder, whose mass is 250 g, when it is supplied with
6000 J of energy. The specific heat capacity of iron is
480 J $kg^{-1} \, K^{-1}$. (3)

H4

Jules sets up an experiment which involves heating up a metal
cylinder, using an immersion heater. He fits the heater into a hole
in the centre of the cylinder and then connects the terminals across
a 12 V supply. The temperature of the metal can be read from the
thermometer.

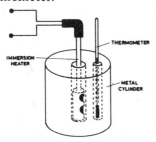

(a) What should Jules do to ensure that very little heat escapes
from the cylinder? (2)

(b) Draw a diagram to show how he should connect a voltmeter
and an ammeter into the circuit, in order to find out the
power supplied by the heater to the metal cylinder. (2)

(c) During one of his experiments, he records the following
information:

      reading on voltmeter = 12.0 V
      reading on ammeter = 2.5 A
      time for which supply is switched on = 5 minutes
      mass of metal cylinder = 1.50 kg
      temperature rise produced = 20°C

(i) What is the power rating of the heater? (2)

(ii) Assuming no loss of heat, how much energy does the
metal receive? (2)

(iii) Calculate the specific heat capacity of this metal. (2)

H5

Graph I shows how a 2.0 kg sample of metal rises in temperature when energy is supplied to it at a rate of 14 000 J min^{-1}
Graph II shows how a 2.0 kg sample of a different metal cools when energy is removed from it at a rate of 14 000 J min^{-1}.

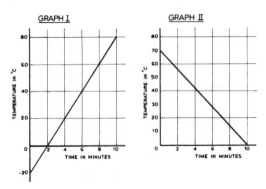

(a) Calculate the specific heat capacity of the metal used for Graph I.   (5)
(b) Calculate the specific heat capacity of the metal used for Graph II.   (5)

H6

Jules's partner, Isa Therm, is looking into what happens when liquids are mixed together. She takes 300 g of water at 20°C and adds it to 200 g of water at 70°C.

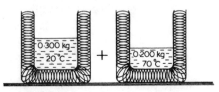

When Isa stirs the mixture thoroughly, a temperature of 40°C is recorded. Assuming that no heat energy is transferred to the containers, answer the following questions.

(a) (i)   What heat energy would be lost by the hot water? The specific heat capacity of water is 4200 J kg^{-1}K^{-1}.   (3)
     (ii) What heat energy would be gained by the cold water?   (3)
(b) She now pours this mixture at 40°C into a beaker containing 200 g of water at 70°C. Find the resulting temperature of this new mixture.   (4)

H7

Ivan Eisberg wants to know how much energy is needed to convert 1 kg of melting ice into water. He takes a supply of ice from the school refrigerator and breaks the ice into small lumps. He dries the lumps and then puts them into two identical plastic funnels. Next, he fits a 12 V, 55 W immersion heater into funnel A, as shown.

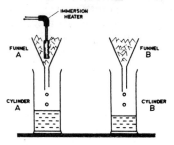

Ivan records the following information:

	Mass of water collected	
	in cylinder A	in cylinder B
5 minutes after start	0.020 kg	0.020 kg
10 minutes after start	0.080 kg	0.040 kg

(a) During the first 5 minutes of his experiment, Ivan does **not** switch on his immersion heater. Explain why the same mass of water collects in each cylinder. (2)

(b) During the next **4 minutes**, the immersion heater is switched **on**. Ten minutes after the start of the experiment, the drip-rate is back to normal. How much ice is melted by the energy from the immersion heater? (2)

(c) How much energy does the immersion heater supply? (2)

(d) From parts (b) and (c), deduce the amount of energy needed to change 1 kg of ice at 0°C into water at 0°C. (3)

(e) What name is given to the quantity calculated in (d)? (1)

H8

(a) The specific latent heat of fusion of ice is 3.3×10^5 J kg^{-1}. How much energy is needed to convert a 2.0 kg block of ice at 0°C into water at 0°C? (4)

(b) A block of ice with a hole dug out of it is placed in a large trough and hot water is poured onto it. The temperature of the water is 99°C and of the ice is 0°C.

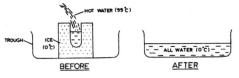

If the mass of the block of ice is 6.3 kg, how much hot water is needed to convert all the ice into water at 0°C? (6)

H9

In the process of 'getting up steam', Jules inserts an immersion heater into an insulated beaker containing water. The beaker rests on a compression balance. The electrical energy supplied to the heater is recorded on a joulemeter. When the water is boiling, Jules zeroes the joulemeter and then records readings from the joulemeter and balance as shown.

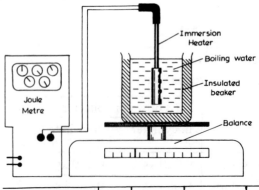

energy supplied in J	0	50 000	100 000	350 000	500 000
reading on balance in kg	1.00	0.98	0.96	0.86	0.80

(a) Draw a graph of the energy supplied against the mass of steam produced. (5)

(b) Deduce from this graph the energy required to convert 1 kg of boiling water entirely into steam. (2)

(c) What name is given to the quantity worked out in (b)? (1)

(d) The value calculated here is slightly higher than that generally quoted. Try to account for this. (2)

H10

Some pupils are using the following apparatus in an experiment to find out how the volume of a sample of gas depends upon the pressure applied to it.

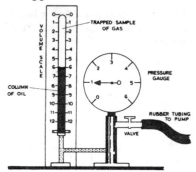

(a) At the start of the experiment, the valve is **open**. Explain why the reading on the gauge is 1 unit, and **not zero**, even although a pump has not yet been connected to the rubber tubing. (1)

(b) Suggest what units must have been used for the scale round the pressure gauge. (1)

(c) The volume of the trapped gas is obviously 6 units at the start. The pupils then connect a pump to the tubing and use it to increase the pressure applied to the gas sample. Part of their table of results is shown:

pressure (units)	1.0	2.0	3.0	4.0	5.0
volume (units)	6.0	?	?	?	?

Fill in the four missing readings and suggest what relationship exists between the pressure and the volume of the gas. (4)

(d) The lowest mark on the volume scale is 12 units. How could the pupils obtain a reading as large as this in practice? What would be the corresponding reading on their pressure gauge? (2)

(e) Why must the temperature of the apparatus be kept constant during their experiment? (1)

(f) What name is given to the relationship deduced from this experiment? (1)

H11

This is a typical experimental set-up used to investigate how the volume of a trapped sample of gas depends upon its temperature (Charles' law).

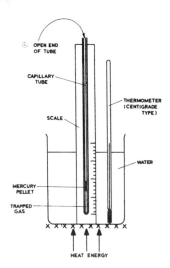

(a) (i) You will notice that the capillary tube used in the experiment has its top end open — why is this necessary? (2)

(ii)   What factor, apart from the mass of the gas, is constant
        during this experiment?   (1)
(b) As the temperature increases, the volume of the trapped gas
     increases in direct proportion ($V \propto T$)
     (i)   How would the volume V be measured with this
            apparatus?   (1)
     (ii)   For V/T to be a constant, the temperature has to be
            measured on the Kelvin scale. Explain how you would
            convert your temperature readings (in °C) into Kelvin
            temperatures.   (2)
(c) The following results were recorded in an experiment, but they
     are incomplete:

| length of trapped air column | 5.8 cm | 6.2 cm | B | 7.4 cm |
| temperature of water | A | 37°C | 77°C | 97°C |

Assuming that the results agree with the law already mentioned,
what are the values of A and B?   (4)

H12
The following apparatus was used during a class experiment to find
out how the pressure of a trapped sample of gas varies with
temperature.

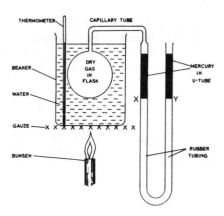

The air in the flask is brought to different temperatures and the
pressure in the flask is deduced from the levels of mercury in tubes
X and Y.
(a) Why must the heat be supplied **slowly** during this experi-
     ment?   (1)
(b) Why is the U-tube always adjusted so that the level of mercury
     in tube X stays the same?   (2)
(c) What would you expect to happen to the mercury level in tube
     Y as the gas is heated up?   (1)
(d) Here is a table of results taken by the class:

I	II	III	IV	V
temperature (in °C)	temperature (in K)	level difference (in cm)	atmospheric pressure (in cm of mercury)	gas pressure (in cm Hg)
20	293	0.0	76.0	76.0
40	313	5.2	76.0	81.2
60	333	10.4	76.0	86.4
80	353	15.6	76.0	91.6
100	?	?	?	?

Draw a graph to find out how the pressure of the gas depends upon its Kelvin temperature. (4)

(e) No readings were taken for 100°C because the boiling water presented difficulties. Using the graph to help you, if necessary, fill in the missing information for columns II to V. (2)

H13

(a) Billy Blaster designs a toy bazooka to fire ping-pong balls at his few remaining friends. He jams a ping-pong ball into the bazooka at A and then pushes in the plunger, which is initially at B.

The pressure of the air inside the bazooka has to reach 4.5×10^5 N m^{-2} (Pa) to force the ball out. Originally, the pressure was 1.0×10^5 N m^{-2} (Pa) and distance BA was 0.90 m. Through what distance must Billy move the plunger to fire the ping-pong ball? (5)

(b) A young skin-diver checks his air cylinder in the comfort of a warm school minibus before entering the chilly waters of the North Sea. He reads the pressure on the gauge and the temperature inside the minibus. When just under the water surface, he records the temperature of the water and waits a few minutes before taking the new reading from the gauge on the cylinder.

(i) What would be the new reading on the pressure gauge, assuming no air had been used up? (3)

(ii) Why did the diver wait a few minutes before reading the gauge? (2)

H14

Billy Blunder has just finished a detailed study of the gas laws and he decides to draw three graphs to summarise the three gas laws. Not for the first time, he makes a few large blunders. Here is a copy of his summary. Redraw each graph the way you think it should be.   (10)

GRAPH I   PRESSURE AGAINST VOLUME

GRAPH II   PRESSURE AGAINST TEMPERATURE

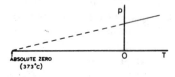

GRAPH III   VOLUME AGAINST TEMPERATURE

Solutions to Worked Examples

H1

(a)   (i)   frictional force at rim = difference in tensions on two sides of belt
               = difference in spring balance readings
               = (120 − 40) N = 80 N

  (ii)   mechanical work done = (frictional force) x (total distance moved against it)
               = 80 x circumference x no. of revolutions
               = 80 x π x 0.08 x 40 = 804 J (approx.)

(b)   The cylinder would heat up. Mechanical energy is being converted into heat energy at the rim of the cylinder as it moves under the friction belt.

(c)   (i)   Insert a thermometer into a hole prepared for it in the cylinder.

  (ii)   The temperature rise is obviously directly proportional to the mechanical energy supplied.

$$\Delta T \propto E \quad \text{or} \quad \frac{E}{\Delta T} = \text{constant}$$

(iii) If a larger mass of metal were used, the temperature rise would be smaller for a given supply of mechanical energy. One would expect different temperature rises when different substances are used.

H2

(a) (i) factors kept constant are: material used (same metal throughout), and the energy supplied to the metal.

(ii) Doubling the mass halves the temperature rise and so the temperature rise is inversely proportional to the mass being heated.

$$\Delta T \propto \frac{1}{m} \quad \text{or} \quad \Delta T \times m = \text{constant}$$

(b) (i) Metal Y.

(ii) Highest specific heat capacity is for metal X because the energy supplied raises its temperature by the smallest amount. Lowest specific heat capacity is for metal Z.

(iii) specific heat capacity $(\text{J kg}^{-1}\,\text{K}^{-1})$

$$= \frac{\text{energy supplied (J)}}{\text{mass heated (kg)} \times \text{temperature rise (}^{\circ}\text{C or K)}}$$

$$c = \frac{E}{m\Delta T} = \frac{2000\ \text{J}}{1.0\ \text{kg} \times 6.0\ \text{K}} = 333\ \text{J kg}^{-1}\,^{\circ}\text{C}^{-1}\ (\text{or J kg}^{-1}\text{K}^{-1})$$

The specific heat capacity of metal Y is $333\ \text{J kg}^{-1}\,\text{K}^{-1}$.

H3

specific heat capacity $(\text{J kg}^{-1}\,\text{K}^{-1}) = \dfrac{\text{energy received (J)}}{\text{mass (kg)} \times \text{temperature rise (K)}}$

(a) $c = \dfrac{E}{m\Delta T} = \dfrac{110\ 000\ \text{J}}{5.0\ \text{kg} \times (43 - 18)\ \text{K}} = 880\ \text{J kg}^{-1}\,\text{K}^{-1}$

(b) $E = cm\Delta T = 385 \times 20 \times (1080 - 80) = 7\ 700\ 000\ \text{J}\ (7.7 \times 10^6\ \text{J})$

(c) $\Delta T = \dfrac{E}{mc} = \dfrac{6000}{0.250 \times 480} = 50\ \text{K}\ (\text{or } 50^{\circ}\text{C})$

H4

(a) Cover the surfaces of the cylinder with insulating material (e.g. expanded polystyrene or cotton wool lagging).

(b)

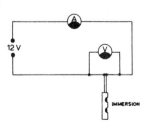

(c) (i)   power rating = (current flowing through heater)
                        x (p.d. across heater)
            $P = IV = 2.5$ A x 12.0 V $= 30$ watts (30 W)
            The rate at which heat energy is supplied is 30 J s^{-1},
            or 30 W.
     (ii) energy supplied = (power rating) x (time of supply)
                $E = Pt = IVt = 30$ W x (5×60) s $= 9000$ J (or 9 kJ)
            The metal receives 9000 J of energy during the experi-
            ment.
     (iii) $c = \dfrac{E}{m\Delta T}$ (as before) $c = \dfrac{9000 \text{ J}}{1.50 \text{ kg} \times 20 \text{ K}} = 300$ J kg^{-1} K^{-1}

            The specific heat capacity of the metal is 300 J kg^{-1} K^{-1}

H5

(a) energy supplied in 10 minutes = 14 000 J min^{-1} x 10 min
                                = 140 000 J
     rise in temperature in this time = $[80 - (-20)] = 100°$C
     (or 100 K)
     specific heat capacity,
     $$c = \frac{E}{m\Delta T} = \frac{140\ 000}{2.0 \times 100} = 700 \text{ J kg}^{-1} \text{ K}^{-1}$$

(b) energy removed in 10 minutes = 14 000 J min^{-1} x 10 min
                                = 140 000 J
     drop in temperature in this time = $70°$C (or 70 K)
     specific heat capacity,
     $$c = \frac{E}{m\Delta T} = \frac{140\ 000}{2.0 \times 70} = 1000 \text{ J kg}^{-1} \text{ K}^{-1}$$

H6

(a) (i)   heat energy lost $E \downarrow = cm\Delta T \downarrow = 4200 \times 0.200 \times (70 - 40)$
                $E \downarrow = 25\ 200 = 2.52 \times 10^4$ J
     (ii) heat energy gained $E \uparrow = cm\Delta T \uparrow = 4200 \times 0.300 \times (40 - 20)$
                $E \uparrow = 25\ 200 = 2.52 \times 10^4$ J
            As expected the energy gained is equal to the energy lost,
            since no heat is lost to the surroundings.
(b) Let $\theta°$C be the final temperature of the new mixture.
     heat energy lost by hot water
            $E \downarrow = c \times 0.200 \times (70 - \theta)$ J
     heat gained by original mixture
            $E \uparrow = c \times (0.300 + 0.200) \times (\theta - 40)$ J
     But
            $E \downarrow = E \uparrow \Rightarrow c \times 0.200 \times (70 - \theta) = c \times 0.500 \times (\theta - 40)$
                        $14 - 0.200\ \theta = 0.500\ \theta - 20$
                                $0.700\ \theta = 34$
                        $\therefore \qquad \theta = 49°$C (approx.)

H7

(a) Since room temperature is more than $0°$C, heat energy is taken
     in by the ice from the room, causing some ice to melt. As

systems A and B are nearly identical, the same mass of water will be collected in each cylinder.

(b) Cylinder A has 0.080 kg after heating whereas cylinder B has only 0.040 kg. The difference, (0.080 − 0.040) kg of water, must have been melted by energy from the immersion heater i.e. **heater melts 0.040 kg of ice.**

(c) Heater supplies energy at a rate of 55 W for (4 × 60) s.
$$E = Pt = 13\ 200\ \text{J}$$

(d) By proportion,
0.040 kg of ice at 0°C is converted into water at 0°C by 13 200 J
1.000 kg of ice at 0°C is converted into water at 0°C by XJ

$$\frac{X\ \text{J}}{13\ 200\ \text{J}} = \frac{1.000\ \text{kg}}{0.040\ \text{kg}}$$

$$X = \frac{13\ 200 \times 1.000}{0.040} = 330\ 000\ \text{J or } 3.3 \times 10^5\ \text{J}$$

3.3 × 10⁵ J of energy are needed to convert 1 kg of ice at 0°C into water at 0°C.

(e) This quantity is the specific latent heat of fusion of ice and is measured in J kg⁻¹.

H8

(a) specific latent heat of fusion (J kg⁻¹) = $\dfrac{\text{energy supplied (J)}}{\text{mass melted (kg)}}$

$$l_f = \frac{E}{m} \Rightarrow E = ml_f = 2.0 \times 3.3 \times 10^5 = 6.6 \times 10^5\ \text{J}$$

(b) Assuming that no heat is exchanged with the surroundings,
energy lost by water cooling = energy needed to change all ice to 0°C   into water at 0°C
$$\Rightarrow (cm\Delta T)_{\text{water}} = m_{ice}\ l_f$$
$$4200 \times m_{\text{water}} \times 99 = 6.3 \times 3.3 \times 10^5$$
$$m_{\text{water}} = \frac{6.3 \times 3.3 \times 10^5}{4200 \times 99} = 5.0\ \text{kg.}$$

H9

(a)

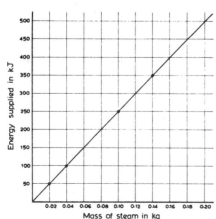

(b)   Since the two quantities are directly proportional, we can say:
energy needed for 1 kg of steam = 10 x energy needed for 0.1 kg
of steam
$$= 10 \times 250\ 000 = 2\ 500\ 000\ J$$
$$= 2.5 \times 10^6\ J.$$

(c)   The energy needed to convert 1 kg of boiling water into steam
is called **the specific latent heat of vaporisation of water** (l_v) and
is measured in J kg^{-1}.

(d)   The high value is due to the fact that energy is lost and so the
joulemeter reading is greater than the energy actually used to
convert the water into steam.

$$l_v = \frac{E}{m} \quad \text{and so if } E \text{ is too high, so is } l_v.$$

H10

(a)   The pressure gauge is reading the pressure of the atmosphere
when the valve is open.

(b)   The pressure of the atmosphere is nearly 1.0×10^5 N m^{-2} (or Pa)
and so the gauge must be calibrated in **atmospheres of pressure**.

(c)

pressure (units)	1.0	2.0	3.0	4.0	5.0
volume (units)	6.0	3.0	2.0	1.5	1.2

The general rule is pressure x volume = constant (6.0 in this
case).
Another way of saying this is that the volume of a fixed mass of
gas is inversely proportional to the pressure applied to it.

(d)   They could use a pump designed to reduce the pressure to
0.5×10^5 Pa. At this pressure the volume would be 12 units
because
pressure x volume = constant.
$0.5 \times 12 = 6.0$

(e)   Any change in temperature would alter the volume of the
trapped gas, introducing another (uncontrolled) factor.

(f)   Boyle's law. It states that the volume of a fixed mass of gas is
inversely proportional to the pressure applied to it, provided the
temperature stays constant.

H11

(a)   (i)   The capillary tube is left open at the top so that the
pellet of mercury is free to move without changing the
pressure of the gas above it, which is normal atmospheric
pressure.

(ii)   The pressure acting on the trapped gas is kept constant
throughout this experiment (almost atmospheric pressure).

(b)   (i)   Since the volume of the trapped gas is directly proportional
to the length of the gas column under the mercury pellet,
length can be measured on the scale and used instead of
volume.

(ii)   The Kelvin temperature scale starts at $-273°C$ (absolute zero) so we convert Celsius temperatures into Kelvin temperatures by simply adding 273.

$$T \text{ (in K)} = 273 + T \text{ (in } °C)$$

e.g. a temperature of 80°C is equivalent to a temperature of (273 + 80) K or 353 K.

(c) Assuming that V/T_K = constant.

$$\frac{6.2}{(273 + 37)} = \frac{7.4}{(273 + 97)} = 0.02$$

Then it follows that

$$\frac{5.8}{A} = 0.02 \Rightarrow A = \frac{5.8}{0.02} = 290 \text{ K (or } 17°C)$$

also

$$\frac{B}{(273 + 77)} = 0.02 \Rightarrow B = 0.02 \times (273 + 77) = 7 \text{ cm}$$

H12

(a) To make sure that the gas is at the same temperature as the water surrounding the flask. If heat were supplied too quickly, the gas would be at a lower temperature than recorded.

(b) This is necessary so that the volume of the trapped gas is kept constant throughout the experiment. Remember that the volume of the gas at the top of tube X and in the capillary tube is being heated as well as that in the flask.

(c) It will rise, relative to the (fixed) level in tube X.

(d) Graph should be a straight line through the origin.

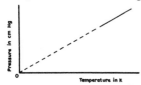

The pressure of a given mass of gas is directly proportional to its Kelvin temperature, provided the volume is kept constant.

(e)

II	III	IV	V
373	20.8	76.0	96.8

H13

(a) Assuming that there is no change of temperature, we can use Boyle's law

$$p_1 V_1 = p_2 V_2 = \text{constant.}$$

Let the new length of the bazooka be l metres and let the cross-sectional area of the tube be A m^2.

$$p_1 V_1 = 1.0 \times 10^5 \times (0.90 \times A)$$

and

$$p_2 V_2 = 4.5 \times 10^5 \times (l \times A)$$

Eliminating the area A and also the figure 10^5 which occurs on both

sides, we are left with
$$0.90 = 4.5 \times l \Rightarrow l = \frac{0.90}{4.5} = 0.20 \text{ m.}$$

New length of the tube is 0.20 m and old length was 0.90 m therefore plunger has been moved in by 0.70 m.

(b) (i)   Assuming that no air escapes from the cylinder and that the volume of the cylinder does not alter, we can use the relationship

$$\frac{p_1}{T_{K_1}} = \frac{p_2}{T_{K_2}} = \text{constant (the subscript } K \text{ refers to Kelvin}$$

temperatures). Filling in the values given, we get

$$\frac{300}{(273 + 27)} = \frac{p_2}{(273 + 7)} \Rightarrow p_2 = \frac{300 \times 280}{300} = 280 \text{ kPa.}$$

The pressure gauge reads 280 kPa at the colder (sea) temperature.

(ii)   The diver waited until the temperature of the air in the cylinder became the same as that of the surrounding water.

H14

GRAPH I

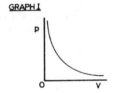

GRAPH II

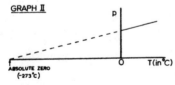

GRAPH III

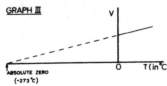

Practice Questions

1. (H1)

The brakes on the back wheel of a bicycle exert an average frictional force of 200 N on the rim when the bicycle is going down a steep hill

The wheel has a diameter of 0.70 m and makes 30 revolutions per minute. How much mechanical work is done against the brake pads per minute?

2. (H2)
(a) What physical quantity has the units J kg^{-1} K^{-1}?
(b) The expression $E = cm\Delta T$ is often used in heat calculations. What do the four symbols represent?
(c) You are told that a certain metal has a high specific heat capacity. Explain what this means about the metal.

3. (H3)
When 5000 J of heat energy are supplied to a metal block of mass 1.00 kg its temperature is raised by 8°C. Calculate the specific heat capacity of the metal.

4. (H3)
How much heat energy is lost to the surroundings when 0.50 kg of water cools down from 80°C to 40°C? (The specific heat capacity of water is 4200 J kg^{-1} K^{-1}.)

5. (H4)
An immersion heater is rated at 25 W. Assuming that the heater is 100% efficient (all the electrical energy converted into heat energy), calculate
(a) the heat energy produced per second
(b) the heat energy produced in 5 minutes.

6. (H4)
An immersion heater is rated at 30 W, but it is only 60% efficient. If this heater is fitted into an aluminium block of mass 1.50 kg and specific heat capacity 880 J kg^{-1} K^{-1}, by how much will it raise the temperature of the block in 5 minutes?

7. (H6)
A piece of metal of mass 0.300 kg at room temperature (20°C) is lowered carefully into a beaker containing 0.200 kg of water at 70°C. The temperature of the mixture becomes 60°C. Assuming that no heat energy is absorbed by the beaker, and that no heat energy is lost to the surroundings, calculate
(a) the heat energy lost by the water.
(b) the heat energy gained by the metal.
(c) the specific heat capacity of the metal.

8. (H7)
Ivan Eisberg swallows a 150 g lump of ice which is at 0°C. How much heat energy would he need to supply to the ice to melt it all? (Take the specific latent heat of fusion of ice to be 3.34 x 10^5 J kg^{-1})

9. (H7, 9)
What heat energy is released when
(i)   150 g of steam at 100°C condenses into water at 100°C?
(ii)   150 g of water at 100°C cools down to 0°C?
(iii)   150 g of water at 0°C changes into ice at 0°C?

Required data
specific latent heat of vaporisation of water = 2.26×10^6 J kg^{-1}
specific latent heat of fusion of ice = 3.34×10^5 J kg^{-1}
specific heat capacity of water = 4.2×10^3 J kg^{-1} K^{-1}

10. (H8, 9)
How much heat energy would be needed to convert 1.20 kg of ice at 0°C into the same mass of steam at 100°C?
(For required data, use values given in previous question)

11. (H10)
Billy Blunder has been doing an experiment to verify **Boyle's law**. His laboratory book shows the correct apparatus — but some strange results as follows.

pressure of gas (in units)	6.0	8.0	11.3	18
volume of gas (in units)	10.0	8.0	6.0	4.0
temperature of gas (in K)	300	320	340	360

(a)   What has Billy Blunder forgotten to keep constant?
(b)   His results do, in fact, show how p, V and T are related for a fixed mass of gas. Can you find the relationship from his results?
(c)   At a temperature of 300 K, what would the volume of the gas be if the pressure on it is 18 units?
(d)   At a temperature of 300 K, what pressure would be needed to produce a volume of 6 units?

12. (H11)
0.14 m^3 of air at 7°C and a pressure of 1.0×10^5 Pa is trapped by a light piston. The temperature of the air is increased by 40°C, without alteration in the pressure. What is the new volume of the air?

13. (H12)
(a)   Draw a graph to show how the pressure of a fixed mass of gas varies with temperature, if its volume is kept constant.
(b)   A fixed mass of gas at a pressure of 1.0×10^5 Pa has its temperature increased from 27°C to 87°C. If there is no change in volume, what would be the new pressure?

14. (H14)
The Kelvin temperature of a fixed mass of gas is doubled and the volume of the gas is reduced to one third of its original value. What can you say about the new pressure of the gas?

15. (H13, 14)
Under standard conditions of temperature and pressure, a gas occupies a volume of 0.10 m³. Find its volume at 27°C and 2 atmospheres of pressure.

16. (H13, 14)
0.40 m³ of gas at 77°C and 2.5 x 10⁵ Pa is brought to standard conditions of temperature and pressure. Find the new volume of the gas.

    standard temperature = 0°C = 273 K
    standard pressure = 1 atmosphere = 1.0 x 10⁵ Pa

UNIT E

Electricity

Content List of Topic Areas

Worked Examples

E1

A polythene rod, charged negatively at end X, is placed on a pivot made from two watch glasses as shown.

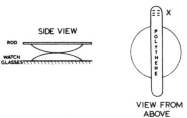

SIDE VIEW

VIEW FROM ABOVE

(a) Describe one method of charging end X negatively. (2)
(b) Explain why the charge at end X does not spread out over the entire polythene rod. (2)
(c) When another rod, charged at end Y, is placed on a nearby pivot, ends X and Y come together. Why is this?

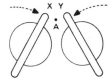

(2)

(d) What would happen if
   (i) the end of another negatively charged rod is positioned at A, between X and Y?;
   (ii) the end of a positively charged rod is positioned at A, between X and Y? (4)

E2

A negatively charged polythene rod is held near to end A of an uncharged metal rod which is mounted on an insulating support.

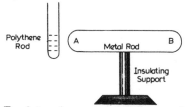

(a) Explain why end A of the metal rod becomes positively charged and end B becomes negatively charged. (4)
(b) Describe how the metal rod could be given a positive charge by the method of induction. Explain this method of charging. (4)
(c) Describe what happens when a positively charged metal rod AB is brought close to the cap of
   (i) a positively charged electroscope.
   (ii) a negatively charged electroscope. (2)

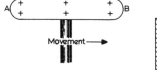

Charged Electroscope

E3

A positively charged acetate rod is held near to two metal balls A and B. The balls are suspended by insulating cords and are initially touching each other.

(a) While the rod is held in the position shown, B is moved away from A. Explain, in terms of the movement of electrons, why B is now positively charged.   (4)

(b) If A is brought close to the cap of a negatively charged electroscope, what would happen to the leaves of the electroscope?   (3)

(c) How could you prove that the charge on A is opposite to the charge on B?   (3)

E4

A physics teacher shows her class an experiment about the electric field effects produced by charged objects. She then leaves the pupils to draw the apparatus. One such attempt is shown below.

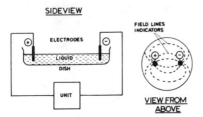

(a) These drawings of the apparatus are not very well labelled.
   (i)   What kind of unit would be used to set up the electric field patterns?
   (ii)   The type of liquid is important. What should it be?
   (iii) What could be used as suitable field line indicators?
   (iv)   No direction has been given to the field lines. What should it be?   (4)

(b) Copy and complete the diagrams that should have appeared in the pupil's report as examples of the electric field patterns.   (6)

(i)   (ii)   (iii)   (iv)

E5

Certain parts of the following passage entitled 'Current and Charge' have been left out. See if you can fill in the missing spaces, using the list below to help you.

List

conductors	ampere	atoms
coulomb	insulators	negatively
charged	second	

Passage

'When an electric current passes through a substance, there is a drift movement of   (a)   particles through the material. Materials through which currents can pass easily are called   (b) , whereas materials through which currents can pass only with great difficulty are called   (c) . Currents pass through metals like copper due to a drift movement of tiny charged particles called electrons. Electrons are   (d)   charged particles and are usually found orbiting the nuclei of   (e) . In circuits, the electrons travel from the negative terminal to the positive terminal.

The basic unit of current is called the   (f) , and the basic unit of charge is called the   (g)   When a current of **one**   (h)   flows for **one**   (i) , a charge of **one**   (j)   passes.'   (10)

E6

The charge Q transferred when a current I flows for a time t is given by $Q = It$.

(a) How much charge passes when
  - (i) a current of 100 amperes (100 A) flows through an ignition circuit of a car for 2 seconds (2 s)?   (1)
  - (ii) a current of 12 A flows through an immersion heater for 15 minutes?   (2)
  - (iii) a current of 16 mA flows from the battery of a transistor for 1 hour?   (2)

The current I is given by $I = Q/t$, using same symbols as above.

(b) What current is flowing in a circuit if
  - (i) 300 coulombs (300 C) of charge are transferred in 1 minute?   (2)
  - (ii) 12 C of charge are transferred in 0.2 s?   (1)
  - (iii) 7.2×10^3 C of charge are transferred in one hour?   (2)

E7

The potential difference, or voltage, across two points is measured by the amount of energy needed to move one coulomb of charge from one point to the other.

$$\text{potential difference (in volts)} = \frac{\text{energy to move charge (in joules)}}{\text{charge moved (in coulombs)}}$$

$$V(V) = \frac{E(J)}{Q(C)}$$

(a)   What is the p.d. (voltage) across the terminals of

   (i)   an immersion heater if 6000 J of heat energy are
         produced when a charge of 500 C passes through its
         element?   (2)

   (ii) a d.c. motor if 2400 J of energy are needed to keep a
         current of 2 A flowing through it for 60 s?   (2)

   (iii) a diode valve if an electron gains 3.2×10^{-16} J of
         kinetic energy as it accelerates from cathode to anode?
         (electronic charge = -1.6×10^{-19} C)   (3)

(b)   How much energy is supplied by the cell, and converted
      into heat in the resistor, in the following circuit?   (3)

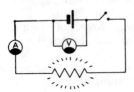

reading on voltmeter = 2.0 V
reading on ammeter = 0.5 A
time of heating = 120 s

E8

The circuit below contains two 1.5 V cells, two identical bulbs and
two identical ammeters.

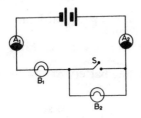

(a)   At the start of an experiment, switch S is **closed** and ammeter
      A_1 registers 0.4 A.

   (i)   Bulb B_1 is shining brightly. What can you say about the
         brightness of bulb B_2? Explain your answer.   (3)

   (ii) What is the reading on ammeter A_2?   (2)

(b)   Switch S is now **opened**.

   (i)   Describe the brightness of the bulbs.   (3)

   (ii) What happens to the readings on the ammeters?   (2)

E9

Two groups of pupils are issued with **identical** sets of apparatus and
asked to find out about the currents in branching circuits. The
circuits they produce and some of the results they obtain are shown
on the next page

Selwyn's group

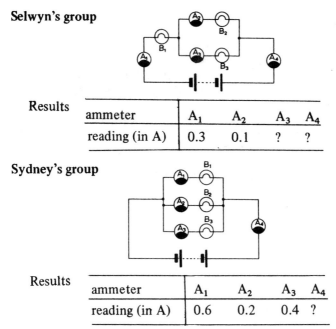

Results

ammeter	A_1	A_2	A_3	A_4
reading (in A)	0.3	0.1	?	?

Sydney's group

Results

ammeter	A_1	A_2	A_3	A_4
reading (in A)	0.6	0.2	0.4	?

Assuming that the resistances of the bulbs do not alter,
(a) what are the correct values of A_3 and A_4 for Selwyn's group? (2)
(b) what is the correct value for A_4 for Sidney's group? (2)
(c) which circuit takes more current from the battery? (2)
(d) which circuit has more resistance? (2)
(e) which bulb has least resistance and which has most? (2)

E10
Billy Blunder, in his normal fashion, has been busy measuring voltages across various groupings of cells and has got the meter readings all mixed up.

CIRCUITS

Meter readings	
(1)	4.5 V
(2)	3.0 V
(3)	1.5 V
(4)	0.0 V
(5)	1.5 V
(6)	3.0 V

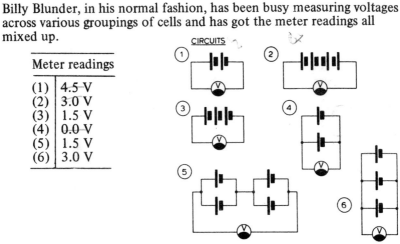

The cells are each 1.5 V and the voltmeter has a very high resistance.

(a) Place the voltmeter readings (1) to (6) so that they fit the circuits ① to ⑥   (6)

(b) Which circuit would supply a 1.5 V bulb with electrical energy at normal brightness for the longest time? Explain your answer.   (2)

(c) Billy Blunder is not at all sure why the needle of a voltmeter deflects. Can you say why?   (2)

E11

Three stages in an experiment are shown in the following circuits.

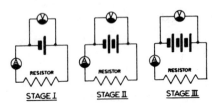

STAGE I   STAGE II   STAGE III

The readings obtained are as follows

	Stage I	Stage II	Stage III
voltmeter reading	2.0 V	4.0 V	6.0 V
ammeter reading	0.5 A	1.0 A	1.5 A

(a) Study the results and suggest a relationship between the current flowing in the circuit and the voltage (p.d.) of the supply.   (2)

(b) What supply voltage is needed to produce a 2.0 A current in this circuit?   (3)

(c) What current would flow through the resistor if a 3.0 V supply were used?   (1)

(d) How would the results change if
   (i)   a higher value,
   (ii) a lower value, of resistance were used?   (2)

(e) According to Ohm's law, the resistance of a resistor is given by

$$\text{resistance (in ohms)} = \frac{\text{p.d. across resistor (in volts)}}{\text{current through resistor (in amperes)}}$$

Calculate the value of the resistor used in the above experiment.   (2)

E12
These problems deal with applications of Ohm's law and each one refers to the same kind of circuit.

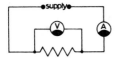

(a) An electric fire is connected to a 240 V mains supply. The resistance of the element is 80 Ω. What current is taken from the supply? (2)

(b) A ray box bulb is to be connected across a 12 V power supply. The supplier states that a current of 2 A flows at this voltage. What is the resistance of the bulb filament? (2)

(c) When all the figures on a calculator are displayed, the current taken from its 9.0 V battery is 20 mA. What is the resistance offered by the components inside to the flow of current? (2)

(d) What battery voltage (p.d.) would be required to cause a current of 100 mA to flow through a resistance of 1.0 k Ω? (2)

(e) A coil of wire has a resistance of 150 Ω. What would be the p.d. across its terminals when the current flowing through the coil is 2 mA? (2)

E13
In the following circuits, each supply is 12.0 V and each resistor has a value of 3.0 Ω. The ammeters have very small values of resistance and the voltmeters have very large values of resistance.

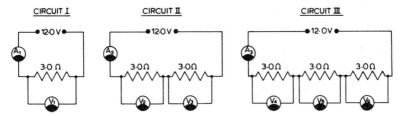

(a) What are the readings on A_1 and V_1? (2)

(b) (i) What is the total resistance of Circuit II? (2)
    (ii) What are the readings on A_2, V_2 and V_3? (3)

(c) (i) What is the total resistance of Circuit III? (1)
    (ii) What are the readings on A_3, V_4, V_5 and V_6? (2)

E14
(a) In the following circuit, V is a high resistance voltmeter and A is an ammeter of negligible resistance.

(i)   What is the purpose of the voltmeter V?   (1)
(ii) What is the purpose of the ammeter A?   (1)
(b) In the circuit below, a voltmeter placed across points A and
B registers 6.0 V. The resistance of the ammeter is, as usual,
negligible and the supply voltage is 12.0 V.

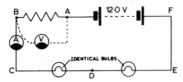

What would the voltmeter readings across the other labelled
parts of the circuit be? Give your answers in table form.

Points in circuit	Reading on voltmeter
A & B	6.0 V
B & C	
C & D	
D & E	
E & F	
F & A	(5)

(c) The bulbs used in the above circuit are each rated '3 V,
0.3 A'. Comment upon their brightness.   (2)
(d) Why would it be inadvisable to short out the resistor AB?   (1)

E 15
The ammeters in the following circuit have negligible resistances and
the voltmeters have very large resistances.

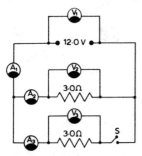

(a) When switch S is **open**, what are the readings on
(i)   the ammeters?
(ii) the voltmeters?   (4)
(b) When S is **closed**, what are the readings on
(i)   the ammeters?
(ii) the voltmeters?   (4)

(c) What is the effective resistance of this circuit when
    (i)   S is open?
    (ii) S is closed?   (2)

E16
Calculate the resistance of each of the following combinations of resistors.

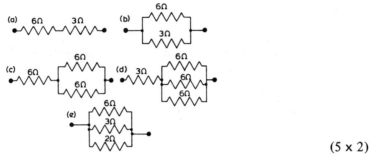

                                                             (5 × 2)

E17
Barbara Blowfuse's hobby is building circuits which baffle her class (and sometimes the teacher!). Here is one of her simpler efforts.

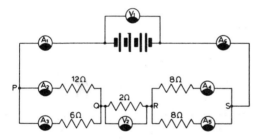

(a) If each cell is 1.5 V, what is the total supply voltage recorded on V_1?   (1)
(b) Calculate the current taken from this supply.   (4)
(c) Deduce the readings on all the ammeters (assumed to be ideal).   (3)
(d) What is the reading on V_2?   (2)

E18
Ivy Watt found three expressions for electrical power in her physics book

    [I] $P = IV$ [II] $P = I^2 R$ [III] $P = \dfrac{V^2}{R}$

(a) Show how she could obtain [II] and [III] from expression [I] by using Ohm's law.   (4)
(b) In an experiment with a small immersion heater, Ivy records the following data.

voltage (p.d.) across the heater = 10 V
current through heater, when hot = 2.5 A

(i)   What is the power rating of this heater?   (2)
(ii) What is the heater element resistance, when hot?   (2)
(iii) How much energy would be transferred by the heater
      in five minutes?   (2)

E19

(a)   A fused three-pin plug of the type commonly used for connect-
      ing electrical equipment to mains power points is shown below.

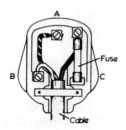

There are three connections in the plug for the three wires in
the cable. The wires are
(i)   the earth wire
(ii) the neutral wire
(iii) the live wire

(i)   Why is it important to have the live wire connected
      first to the fuse and then to pin C?   (2)
(ii) To which pin (A or B) is the earth wire connected?   (1)
(iii) What is the modern colour code for identifying the
      three wires?   (2)

(b)   This is a diagram of a ring circuit wiring system, commonly
      used in houses. The earth wire has been omitted to keep the
      diagram simple. Appliances which could be connected to power
      points in the ring circuit are represented by resistors A, B and C.

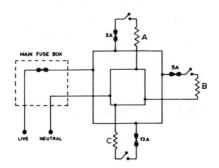

(i) Explain why the switches are placed in the live line between the fuse and the appliance. (2)

(ii) What advantage is there in having each plug which fits into the system separately fused? (2)

(iii) Give one example of an appliance which could be used in position A. (1)

E20

A brass rod, p q, is held within the magnetic field of two large magnets as shown. The ends of the rod are soldered to two long pieces of connecting wire, ap and bq.

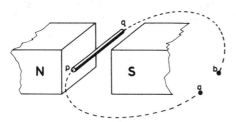

(a) Draw a diagram to indicate the magnetic field pattern between the two magnetic poles. (2)

(b) Describe what you would expect to happen to the rod pq if a d.c. supply is connected into the gap ab as follows:

(2)

(c) Instead of the cell a sensitive centre-zero galvanometer (ammeter) is connected between a and b. What would you have to do to the rod to produce a current reading on this galvanometer? (2)

(d) The effects in part (b) obey the **motor rule** and the effects in part (c) obey the **generator rule**. Give an outline of the two rules. (4)

E21

This is a diagram of a model motor/generator which can be built up from a kit. As you can see, the parts of the assembly have not been labelled.

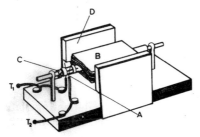

(a) What additional piece of equipment would be required to make it a working motor and where would it be fitted? (2)

(b) Name the parts which are labelled A, B, C and D and state the function of each in the model **motor**. (4)

(c) What could be done to increase the speed of rotation of the motor? (1)

(d) What alterations, if any, are required to make this model act as a **generator**? Describe how the generator works. (3)

E22

Two coils of copper wire are wrapped round hollow cylinders of cardboard and then placed near each other on a bench as shown, G is a sensitive, centre-zero galvanometer and is used to detect any current flowing in circuit B .

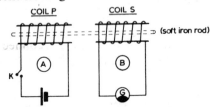

(a) There is no reading on G while switch K is open or while it is closed BUT there are current readings on G while K is **being** closed and while K is **being** opened. Explain this in terms of magnetic field effects. (4)

(b) Why must the galvanometer used be a centre-zero type? (2)

(c) What could be done in circuit Ⓐ to increase the currents induced in circuit Ⓑ ? (2)

(d) If a long rod of soft iron is used to couple the coils and the switch K is then opened or closed, the reading on G increases dramatically. Why? (2)

E23

A simplified diagram of the X and Y deflecting system, used in an oscilloscope, is shown below. At the start of an experiment, the spot produced on the fluorescent screen by the electron beam is at position O.

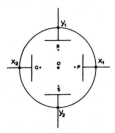

(a) By applying suitable d.c. voltages across the deflecting plates X_1 X_2 and Y_1 Y_2, the spot can be deflected to other parts

of the screen. How would you deflect the spot to
(i)   position P,
(ii) position Q,
(iii) position R,
(iv) position S? (4)
(b) An alternating voltage (a.c. signal) is applied across the Y-plates. What would you expect to see on the screen if the frequency is
(i)   1 Hz, and
(ii) 100 Hz? (2)
(c) When the time-base circuit is applied to the X-plates, the electron beam is made to scan the screen from left to right, return rapidly and repeat the scan. What will be seen on the screen if the time base frequency is
(i)   1 Hz, and
(ii) 100 Hz? (2)
(d) When the time-base is operating and an a.c. supply is connected across the Y-plates, the following trace appears on the screen:

Sketch the wave pattern you would expect when
(i)   the time base frequency is doubled
(ii) the Y-gain control is turned up (increased gain). (2)

E24
This is a diagram of a thermionic diode valve which is shown connected to an a.c. mains supply through a suitably designed transformer.

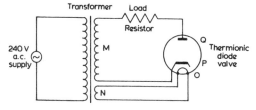

(a) O, P and Q are all involved in obtaining current flow through the valve.
(i)   Name the parts O, P and Q.
(ii) Describe briefly their main functions in the valve itself. (6)
(b) Parts M and N are concerned with the power supplied to the valve.
(i)   Why are two separate power supplies, M and N, required?
(ii) One of the supplies must be limited to a low, fixed value. Which one would this be (M or N), and why must the recommended voltage value be strictly observed? (4)

E25

The thermionic diode and the semiconductor diode are both types
of RECTIFIER. The diode in the diagram can be assumed to
represent, in action, either of the two types.

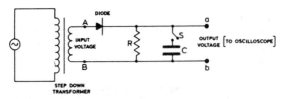

(a)   What is meant by the term 'rectification'?   (1)
(b)   Sketch the waveform which would be seen on a suitably
       adjusted oscilloscope with its time base on, when switch S is
       (i)   open,
       (ii) closed.   (3)
(c)   Describe briefly the main function of the capacitor C.   (2)
(d)   The rectification described above is called 'half-wave
       rectification'. What modifications would have to be made to
       the circuit to obtain a fully rectified output?   (4)

Solutions to Worked Examples

E1

(a)   End X could be charged by rubbing with a material such as
       flannel. A small amount of rubbing can produce a large number
       of charges (ions).
(b)   Polythene is a good insulator so there will be no significant flow
       of charge through it.
(c)   The fact that both rods are charged and come together suggests
       that they are **oppositely** charged. Since X is negative, Y must be
       positive because **unlike charges attract each other.**
(d)   (i)   X would be repelled from A (like charges repel each other)
             Y would be attracted towards A (unlike charges attract)
       (ii) X would be attracted towards A, and Y would be repelled
             from A.

E2

(a)   In the metal rod, electrons are repelled by the negatively charged
       polythene rod and they move from end A along to end B. Since
       electrons are tiny, negatively charged particles, end B becomes
       negatively charged. End A has obviously lost electrons and so
       becomes positively charged.

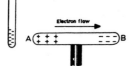

(b) If, now, end B is earthed by touching it with a finger, electrons travel down to earth to get as far away as possible from the repulsion of the polythene rod. End A is still positive, but end B has become neutral.

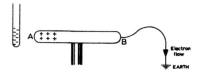

The earth connection is removed and then the polythene rod is removed. The positive charge on A spreads over the metal surface.

This method of charging is called **induction**.

(c) Case (i) The electroscope leaves move farther apart (diverge).
Case (ii) The electroscope leaves come closer together (converge).

E3

(a) Being tiny negatively charged particles, electrons are attracted from B onto A by the positively charged rod. Ball A becomes negatively charged (surplus of electrons) and B becomes positive (shortage of electrons). If B is moved away, the two types of charge have been separated out by induction and B is positively charged.

(b) Since A is now negatively charged, it will make the leaves of the negatively charged electroscope go farther apart.

(c) Bring each in turn close to the electroscope. The leaves will diverge when A is near, but converge when B is near. The balls thus have opposite charges.

E4

(a) (i) a high voltage d.c. supply capable of delivering a few kilovolts

(ii) a non-conducting liquid e.g. a light oil

(iii) grass seeds, small bits of cork, semolina, etc.

(iv) The field direction is defined as the direction in which a positive test charge would be moved in the field. The field is directed from + to −. ⊙ ●--→--● ⊖

(b)

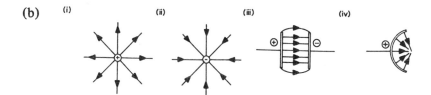

E5
(a)   charged   (b)   conductors   (c)   insulators   (d)   negatively
(e)   atoms   (f)   ampere   (g)   coulomb   (h)   ampere
(i)   second   (j)   coulomb.

E6
(a)   (i) $Q = It = 100$ A x 2 s $= 200$ C
(ii) $Q = It = 12$ A x (15×60) s $= 10\ 800$ C
(iii) $Q = It = (16 \times 10^{-3})$ A x (60×60) s $= 57.6$ C

(b)   (i) $I = \dfrac{Q}{t} = \dfrac{300 \text{ C}}{60 \text{ s}} = 5$ A

(ii) $I = \dfrac{Q}{t} = \dfrac{12 \text{ C}}{0.2 \text{ s}} = 60$ A

(iii) $I = \dfrac{7.2 \times 10^3 \text{ C}}{3.6 \times 10^3 \text{ s}} = 2$ A

E7
(a)   (i) $V = \dfrac{E}{Q} = \dfrac{6000 \text{ J}}{500 \text{ C}} = 12$ V

(ii) $V = \dfrac{E}{Q} = \dfrac{2400 \text{ J}}{(2 \times 60) \text{ C}} = 20$ V

(iii) $V = \dfrac{3.2 \times 10^{-16} \text{ J}}{1.6 \times 10^{-19} \text{ C}} = 2.0 \times 10^3$ V (or 2 kV)

(b) $E = QV = ItV = 0.5$ A x 120 s x 2.0 V $= 120$ J

E8
(a)   (i) B_2 will not be lit because there is a short circuit across it when S is closed. All the current will take the route through the switch because S has effectively zero resistance compared with the bulb.
(ii) $A_2 = A_1 = 0.4$ A. The current is the same in all parts of a series circuit.
(b)   (i)   When S is open the only route for current is through both bulbs. The resistance of the circuit is increased and so the current flowing will be smaller and the bulbs will be quite dim.
(ii) A_1 and A_2 readings will drop to the same, lower value (approximately 0.2 A).

E9
(a)   **Selwyn's group** $A_3 = 0.2$ A because the sum of the branch currents must equal the current in the main circuit ($A_1 = A_2 + A_3$ $0.3 = 0.1 + 0.2$). Also $A_4 = 0.3$ A because the current is the same in all parts of the main circuit ($A_1 = A_4 = 0.3$).
(b)   **Sidney's group** $A_4 = 1.2$ A, because the current in the main

circuit is equal to the sum of the currents in the three branches
($A_4 = A_1 + A_2 + A_3$).

(c) Sidney's circuit takes more current, 1.2 A as opposed to 0.3 A.

(d) Selwyn's circuit has more resistance as it takes less current from
the battery.

(e) Looking at the results in Sidney's circuit, we see that B_1 has the
largest current flowing through it and so it has the least
resistance. B_2 has the smallest current, so it has the most
resistance.

E10

(a)

circuit number	1	2	3	4	5	6
voltmeter reading	3.0 V	0.0 V	4.5 V	1.5 V	3.0 V	1.5 V

(b) Circuit 6 acts as the best supply. Cells in parallel do not
increase the voltage available at the terminals (still 1.5 V), but
they are able to supply the same current for a longer time
(in this case, for about 3 times as long as a single cell).

(c) A voltmeter is essentially a very sensitive current meter with a
very large resistance in series with it. This means that it takes
very little current from the circuit and so does not disturb
the state of the circuit when placed across a component.

A voltmeter is a **high resistance meter and is placed in parallel**
whereas an ammeter is a **low resistance meter and is placed
in series**.

E11

(a) From the results, doubling the voltage produces double the
current and trebling the voltage produces treble the current.
This suggests that the current is directly proportional to the
voltage. ($I \propto V$)

(b) By proportion, $V = 8.0$ V when $I = 2.0$ A.

(c) By proportion, $I = 0.75$ A when $V = 3.0$ V.

(d) (i) A higher value of resistor means **more opposition** to
current flow and so the current values would all **reduce**.

   (ii) A lower value of resistor means **less opposition** to current
flow and so the current values would all **increase**.

(e) $R = \dfrac{V}{I} = \dfrac{2.0 \text{ V}}{0.5 \text{ A}} = 4.0 \ \Omega$

E12

The problems involve the formula $R = \dfrac{V}{I}$ or different forms of it.

$V = IR$ or $I = \dfrac{V}{R}$ (try to remember these three forms)

(a)   Current wanted $I = \dfrac{V}{R} = \dfrac{240 \text{ V}}{80 \text{ } \Omega} = 3 \text{ A}$

(b)   Resistance wanted $R = \dfrac{V}{I} = \dfrac{12 \text{ V}}{2 \text{ A}} = 6 \text{ } \Omega$

(c)   Resistance wanted $R = \dfrac{V}{I} = \dfrac{9.0 \text{ V}}{0.020 \text{ A}} = 450 \text{ } \Omega$

(d)   Voltage wanted $V = IR = 0.100 \text{ A} \times 1000 \text{ } \Omega = 100 \text{ V}$

(e)   P.D. wanted $V = IR = 0.002 \text{ A} \times 150 \text{ } \Omega = 0.3 \text{ V}$

E13

Let V_s stand for the supply voltage and R_t stand for the total circuit resistance.

(a)   By Ohm's law, current $I_1 = \dfrac{V_s}{R_t} = \dfrac{12.0 \text{ V}}{3.0 \text{ } \Omega} = 4.0 \text{ A}$

$$V_1 = V_s = 12.0 \text{ V}$$

(b)

(i) $R_t = 3.0 \text{ } \Omega + 3.0 \text{ } \Omega = 6.0 \text{ } \Omega$ (resistors in series add together)

(ii) $I_2 = \dfrac{V_s}{R_t} = \dfrac{12.0 \text{ V}}{6.0 \text{ } \Omega} = 2.0 \text{ A}$

By Ohm's law $V_2 = I_2 \times 3.0 \text{ } \Omega = 2.0 \text{ A} \times 3.0 \text{ } \Omega = 6.0 \text{ V}$

Similarly, $V_3 = 6.0 \text{ V}$ (half supply voltage across each resistor)

(c)

(i) $R_t = 9.0 \text{ } \Omega$

(ii) $I_3 = \dfrac{12.0 \text{ V}}{9.0 \text{ } \Omega} = 1\frac{1}{3} \text{ A}, V_4 = I_3 \times 3.0 = 4.0 \text{ V}$

Similarly $V_5 = 4.0 \text{ V}$ and $V_6 = 4.0 \text{ V}$ (one third of supply voltage across each resistor)

I   Resistors in series add together.

II   The sum of the p.d.s (voltages) across a set of series resistors is equal to the p.d. of the supply.

E14

(a)   (i)   The voltmeter measures the potential difference (p.d.) across the resistor. The p.d. is measured in **volts**.

(ii)   The ammeter measures the current that is flowing in the circuit. Current is measured in **amperes**.

(b)   To solve this problem, it is important to keep in mind that

I   the sum of the p.d.s around a series circuit is equal to the supply voltage

II   p.d.s only arise across components which have resistance.

Neither the ammeter nor the switch has resistance and so the p.d. across them is zero. The resistor and each of the bulbs have resistance and so p.d.s will develop across them.

Points in circuit	Reading on voltmeter
A & B	6.0 V (given)
B & C	0.0 V (ammeter has zero resistance)
C & D	3.0 V (bulb has resistance)
D & E	3.0 V (bulb has resistance)
E & F	0.0 V (connecting wire has zero resistance)
F & A	12.0 V (p.d. across the battery)

Note: 12.0 V = 6.0 V + 0.0 V + 3.0 V + 3.0 V + 0.0 V

(c) The correct voltage of 3.0 V has developed across each bulb and so they light up to recommended brightness.

(d) With R shorted out, the full 12.0 V would be across two bulbs. This means that the p.d. across one bulb is 6.0 V i.e. **double the recommended voltage**. The bulb filaments would probably burn out.

E15

(a) Only the top branch of the circuit is in action when S is open and so the total circuit resistance is $R_t = 3.0\ \Omega$.

$$\text{current from supply} = I = \frac{V_s}{R_t} = \frac{12.0\ V}{3.0\ \Omega} = 4.0\ A$$

Readings on A_1 and A_2 are 4.0 A but reading on A_3 is zero since S open.
Reading on $V_1 = V_s = 12.0\ V$ (V_1 is across the supply).
Reading on $V_2 = IR = 4.0\ A \times 3.0\ \Omega = 12.0\ V$.
Reading on $V_3 = 0\ V$ since S is open and no current flows through bottom branch.

Summary

A_1	A_2	A_3	V_1	V_2	V_3
4.0 A	4.0 A	0 A	12.0 V	12.0 V	0 V

(b) Both branches of the circuit are now in action when S is closed. The 12.0 V supply acts **both across the top branch and across the bottom branch** ∴ $V_1 = 12.0\ V$, $V_2 = 12.0\ V$, $V_3 = 12.0\ V$

$$\text{current in top branch} = I_2 = \frac{V_2}{R} = \frac{12.0\ V}{3.0\ \Omega} = 4.0\ A$$

$$\text{current in bottom branch} = I_3 = \frac{V_3}{R} = \frac{12.0\ V}{3.0\ \Omega} = 4.0\ A$$

total current from supply $= I_1 = I_2 + I_3 = 4.0 + 4.0 = 8.0$ A

Summary

A_1	A_2	A_3	V_1	V_2	V_3
8.0 A	4.0 A	4.0 A	12.0 V	12.0 V	12.0 V

└──common voltage──┘

(c)

(i)　$R_t = 3.0\ \Omega$ since only top branch is in action.

(ii)　effective resistance $= \dfrac{\text{supply voltage}}{\text{supply current}} = \dfrac{12.0\text{ V}}{8.0\text{ A}} = 1.5\ \Omega$

Two 3.0 Ω resistors in parallel act like a 1.5 Ω resistor.

E16

To combine resistors in **series**, use the relationship

$$R_s = R_1 + R_2 + R_3 \ldots \text{etc.}$$

To combine resistors in **parallel**, use the relationship

$$\frac{1}{R_p} = \frac{1}{R_1} + \frac{1}{R_2} + \frac{1}{R_3} \ldots \text{etc.}$$

(a)　This is a series combination. $R_s = 6\ \Omega + 3\ \Omega = 9\ \Omega$

(b)　This is a parallel combination.

$$\frac{1}{R_p} = \frac{1}{6} + \frac{1}{3} = \frac{3}{6} \Rightarrow R_p = \frac{6}{3} = 2\ \Omega$$

(c)　This is a (series + parallel) combination.
$R_t = (6 + R_p)$

$$\frac{1}{R_p} = \frac{1}{6} + \frac{1}{6} = \frac{2}{6} \Rightarrow R_p = \frac{6}{2} = 3\ \Omega$$

$$R_t = 6 + 3 = 9\ \Omega$$

(d)　As in part (c) except we have three resistors in parallel

Hence $R_t = 3 + 2 = 5\ \Omega$

(e)　This is a parallel set of three resistors.

$$\frac{1}{R_p} = \frac{1}{6} + \frac{1}{3} + \frac{1}{2} = \frac{12}{12} = 1 \Rightarrow R_p = 1\ \Omega$$

E17

(a) The cell on the right of the supply is connected the wrong way round and so cancels out the voltage of one other cell, leaving only **two cells** effective ∴ the supply voltage $V_s = 2 \times 1.5 = 3.0$ V.

(b) $\dfrac{1}{R_{PQ}} = \dfrac{1}{12} + \dfrac{1}{6} = \dfrac{3}{12} \Rightarrow R_{PQ} = \dfrac{12}{3} = 4\ \Omega$

$\dfrac{1}{R_{RS}} = \dfrac{1}{8} + \dfrac{1}{8} = \dfrac{1}{4} \Rightarrow R_{RS} = \dfrac{4}{1} = 4\ \Omega$

$R_t = R_{PQ} + R_{QR} + R_{RS} = 4 + 2 + 4 = 10\ \Omega$

Supply current $I_s = \dfrac{V_s}{R_t} = \dfrac{3.0 \text{ V}}{10\ \Omega} = 0.3$ A.

(c) Since supply current is 0.3 A, readings on A_1 and A_6 are each 0.3 A. In branching section PQ, the current splits up in the ratio 1:2.
Reading on A_2 is 0.1 A and reading on A_3 is 0.2 A.
In branching section RS, the current splits equally.
Reading on A_4 is 0.15 A and reading on A_5 is 0.15 A.

Summary

A_1	A_2	A_3	A_4	A_5	A_6
0.3 A	0.1 A	0.2 A	0.15 A	0.15 A	0.3 A

(d) Reading on $V_2 = V_{QR} = I_s R_{QR} = 0.3$ A $\times$ 2.0 Ω = 0.6 V

E18

(a) $P = IV$, but $V = IR$ by Ohm's law
$\Rightarrow P = I \times (IR) = I^2 R$

also by Ohm's law, $I = \dfrac{V}{R}$ and so

$P = IV = \left(\dfrac{V}{R}\right) \times V = \dfrac{V^2}{R}$

(b) (i) $P = IV = 2.5$ A $\times$ 10 V = 25 watts (25 W)

(ii) $R = \dfrac{V}{I} = \dfrac{10 \text{ V}}{2.5 \text{ A}} = 4\ \Omega$

(iii) energy = power $\times$ time ∴ $E = IVt$
$E = 2.5$ A $\times$ 10 V $\times$ (5 $\times$ 60) s = 7 500 J (7.5 kJ)

E19

(a) (i)   If the current exceeds the rating of the fuse, due perhaps
to a fault in the appliance, the fuse wire would 'blow' and
break the electrical circuit. The appliance has become
completely disconnected from the dangerous live wire and
so no harm will come to anyone handling the faulty fused
appliance.

(ii)   The earth wire is connected to pin A.

(iii) Code

earth (pin A) – yellow/green
neutral (pin B) – blue
live (pin C) – brown

(b) (i)   When the switch is open, the appliance itself will be isolated
from the mains supply live cable, making it much safer,
especially if the appliance has exposed electrical connections
e.g. a bar heater fire.

(ii)   Separately fusing the plugs means that only the appliance
connected to the plug would stop working if a fault
developed in it. The rest of the circuit would not be affected.

(iii) Appliance A is fused at 2 A. This means that the appliance
must have a power rating less than 500 W e.g. television,
table lamp, cassette recorder, etc.

E20

(a)   Magnetic field pattern

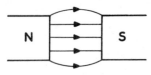

(b) (i)   Rod pq would be forced vertically upwards, at right
angles to the field lines.

(ii)   Rod pq would be forced vertically downwards, at right
angles to the field lines (**motor effect**)

(c)   Rod pq would have to be moved rapidly upwards or rapidly
downwards, at right angles to the field lines (**induced current
effect**)

(d)   Motor rule (right-hand rule)
If first finger points N to S along the field lines, and if centre
finger points in the direction of electron flow in the wire, then
the thumb points in the direction of the resulting motion.

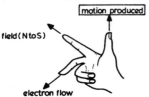

Generator rule (left-hand rule)
If first finger points N to S along the field lines, and if thumb
points in the direction of motion of the wire, then centre finger
points in the direction of the resulting induced electron flow in
the wire.

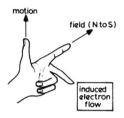

E21

(a) For a motor, a power supply (battery) would be needed. It
must be a d.c. supply to get this model to operate. The
terminals of the battery should be connected to T_1 and T_2.

(b) A — Brushes (these are the electrical connections to the coil)

B — Armature or Coil (the current flowing in this coil
interacts with the magnetic field to produce the motor
force)

C — Commutator (for a simple motor, this would be a split-ring
which reverses the current in the coil every $\frac{1}{2}$-cycle and so
allows the rotation to continue)

D — Magnet (produces a magnetic field necessary for the motor
effect)

(c) The armature would spin faster by having

(i)   more turns on the coil

(ii) more current through the coil (larger battery voltage)

(iii) stronger magnets to make a stronger magnetic field.

(d) The generator is basically a motor in reverse. No battery is
required. The armature has to be rotated mechanically and the
generated voltage can be measured at T_1 and T_2.

E22

(a) When K is open, no current flows in coil P and so there is no
magnetic field around it. When K is closed, current flows in
coil P and creates a steady magnetic field. This field cuts
through the turns of coil S.

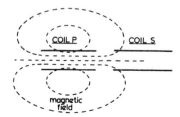

A current will be induced in coil S whenever the magnetic field cutting it **moves** or **changes value**. Obviously, the field does move when K is being closed (field grows outwards from coil P) and also when K is being opened (field shrinks inwards towards coil P).

Currents are induced when magnetic fields are on the move.

(b)   When K is being closed, induced current flows one way through G. When K is being opened, induced current flows the opposite way through G. By having a centre-zero, the needle can deflect in either direction as required.

(c)   To increase the currents induced in circuit B, coil P must be made into a stronger electromagnet, either by increasing the number of turns on it, or by using a larger value of battery voltage. More rapid switching action would also help because **the induced current is proportional to the RATE at which the magnetic field changes.**

(d)   The rod serves to strengthen the electromagnet and increase the magnetic coupling between the coils.

E23

(a)   (i)   Apply a d.c. voltage with X_1 positive and X_2 negative, thereby forcing the electron beam towards X_1 (electrons are negatively charged).

(ii)   Apply d.c. voltage with X_2 positive and X_1 negative.

(iii) Make Y_1 positive and Y_2 negative, by applying a d.c. voltage.

(iv)   Make Y_2 positive and Y_1 negative, by applying a d.c. voltage.

Electrons are deflected towards the more positive plate.

(b)   (i)   The spot would oscillate slowly (1 Hz) in the y-direction.

(ii)   The spot would oscillate so rapidly (at 100 Hz) that a vertical line would appear.

(c)   (i)   The spot would scan in the x-direction at 1 Hz.

(ii)   A horizontal line would appear on the screen at 100 Hz.

(d)   (i)   Since the time of scan is now half its previous value, only ONE wave would be formed on the screen instead of two.

(ii)   The amplitude of the original wave would increase.

E24
(a)   (i)   O is the heater filament, P is the cathode, Q is the anode.
       (ii)   Radiation from the heated filament O raises the temperature of the cathode P and it emits electrons. If the anode Q is made positive in voltage, the thermionic electrons will be attracted through the vacuum inside the valve, from cathode (−) to anode (+). Current can flow from P to Q, but not from Q to P. If the anode is made negative, no electrons will be attracted across to it and so no current will flow. This explains why the device permits flow of current in one direction only − hence the term 'valve'.
(b)   (i)   Power supply N is used to heat up the filament O inside the valve, so that the nearby cathode P can emit electrons. Power supply M is needed to drive electrons through the valve from P to Q. Conduction occurs when this supply makes Q more positive than P.
       (ii)   The voltage recommended by the manufacturers for the filament supply N must be used. Too large a value would result in a burnt out filament. Too low a value would result in a poor supply of thermionic electrons.

E25
(a)   Rectification is the name given to the changing of an a.c. input voltage into a d.c. output voltage. A device which does this is called a rectifier.
(b)   (i)   **S open**

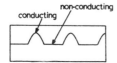

The output voltage waveform shows half-wave rectification. The diode only conducts when end A of the input is more positive than end B, i.e. it only conducts for half of each input cycle. When B is more positive than A, current cannot flow through the diode because it is reverse-biased.

(ii) **S closed**

Here the capacitor C has smoothed the output taken over resistor R. Notice that new output is less pulsating than the original (shown dotted).

(c) The function of the capacitor is to smooth the output voltage of the circuit. It charges up through the diode on the conducting half-cycle and discharges through R (the load) on the non-conducting half-cycle. The slight variation of output voltage that still exists is called the ripple voltage.

(d) A suitable full-wave rectifier must allow for conduction on both half-cycles of the a.c. input at AB. The most common type used is the bridge rectifier shown below.

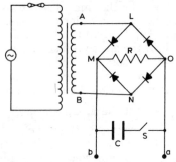

When A is more positive than B, the electrons flow in path BNO(R)MLA. When A is more negative than B, the electrons flow in path ALO(R)MNB. Since electrons flow through R in direction OM on both half-cycles, a full-wave output is obtained across ab

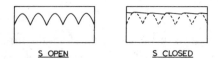

Practice Questions

1. (E1)
Fill in the missing words for each statement below:
(a) Negative charges attract _____ charges.
(b) Positive charges _____ positive charges.

(c)   Negative charges _____ negative charges.
(d)   Positive charges attract _____ charges.

2. (E2)
A metal rod is fitted on top of an insulating base support. How could you charge this metal rod **positively**, by using a **negatively** charged polythene rod?

3. (E3)
An electroscope is given a positive charge. What happens to the leaves of the electroscope when
(a)   a positively charged acetate rod is brought near to its cap?
(b)   a negatively charged polythene rod is brought near to its cap?

4. (E3)
What are the two missing words, (a) and (b), in the following statements?
(a)   Negative charge is usually due to a _____ of electrons.
(b)   Positive charge is usually due to a _____ of electrons.

5. (E4)
Draw diagrams to represent the electric fields you would expect to find around each of the following:
(a)   a small sphere which is positively charged;
(b)   a small sphere which is negatively charged;
(c)   two parallel metal plates, one charged positively, the other charged negatively.

6. (E6)
In which of the following examples is most charge transferred?
(a)   A current of 0.2 A flows through a light bulb for 1 minute.
(b)   A current of 10 mA flows from a transistor battery for 15 minutes.
(c)   A current of 20 A flows through a large d.c. motor for 0.5 s.

7. (E6)
(a)   What current is flowing in a circuit if 120 C of charge are transferred each minute?
(b)   The electronic charge is 1.6×10^{-19} C. If 6.25×10^{18} electrons pass through a circuit each second, what is the current flowing?

8. (E7)
What is the p.d. (voltage) across the terminals of each of the following?
(a)   an electric bulb, if 4800 J of heat and light energy are produced when 20 C of charge pass through the filament
(b)   an immersion heater, if 4.6×10^4 J of energy are produced when a current of 4 A flows through it for 100 s

9. (E7)
During an experiment, an immersion heater delivers 7200 J of energy when operated from a 12 V supply for 5 minutes. Assuming 100% efficiency, what current must flow through the element of this heater?

10. (E8)
Twenty Christmas tree bulbs are connected in **series** across a 240 V supply.
(a)   What can you say about the size of the current which flows through each bulb?
(b)   One of the bulbs gets smashed and all the others go out. The circuit is fixed temporarily by simply shorting out the smashed bulb. Explain the effect this repair would have on the brightness of the remaining bulbs.

11. (E9)
Four bulbs are connected in **parallel** across a suitable d.c. supply. The bulbs are identical.
(a)   If the current delivered by the supply is known to be 8 A, what can you say about the size of the current flowing through each bulb?
(b)   If one of the bulbs gets smashed, explain the effect upon the brightnesses of the remaining three bulbs.

12. (E9)
Fill in the missing spaces in the following table of results, taken from experiments on branching circuits.

Experiment	Current in main circuit	Current in top branch	Current in middle branch	Current in bottom branch
A	6.0 A	2.0 A	1.0 A	(a) ?
B	4.5 A	1.1 A	(b) ?	2.6 A
C	8.0 A	(c) ?	4.0 A	1.5 A
D	(d) ?	0.6 A	0.5 A	0.4 A

13. (E9)
For each of the experiments in 12. (E9), state which branch (top, middle, or bottom) has the greatest resistance.

14. (E10)
(a)   Pupils are given two identical 2.0 V cells and are asked to connect them both so that they give the following voltages:
(i)   4.0 V   (ii)   2.0 V
Describe how the cells should be connected in each case.
(b)   A 9.0 V battery is constructed by connecting a number of 1.5 V cells in series. How many cells would be needed?

15. (E10)
In selecting suitable electric meters, what should you always keep in mind about the resistances of
(i) an ammeter?
(ii) a voltmeter?

16. (E11)
The results of an Ohm's law experiment are shown below. Two of the readings have been omitted.

p.d. across resistor	2.0 V	4.0 V	6.0 V	8.0 V	10.0 V
current through resistor	(x) ?	1.0 A	1.5 A	2.0 A	(y) ?

(a) What is the resistance of the resistor in ohms?
(b) What are the missing readings (x) and (y)?
(c) If a 3.0 V supply were applied across the resistor, what current would flow?

17. (E12)
Use Ohm's law to help you fill in the missing readings (a) to (d).

Appliance	Voltage	Current	Resistance
electric fire	240 V	3.0 A	(a) ?
electric cooker	240 V	(b) ?	16.0 Ω
electric bulb	(c) ?	2.0 A	6.0 Ω
transistor radio	9.0 V	0.045 A	(d) ?

18. (E13)
Resistors of value 2.0 Ω and 4.0 Ω are connected in series to a 12 V d.c. supply.
(a) What current flows through the resistors?
(b) What p.d. develops across the
   (i) 2.0 Ω resistor?
   (ii) 4.0 Ω resistor?

19. (E13)
Twenty christmas tree lights are connected in series across a 240 V supply. The operating resistance of each bulb is 120 Ω.
(a) What is the total resistance of the circuit?
(b) What p.d. (voltage) develops across each bulb?
(c) What current flows through each bulb?

20. (E14)
Four resistors valued 1.0 Ω, 3.0 Ω, 4.0 Ω and 8.0 Ω are connected in series across a 16 V supply.
(a) What voltage would you expect across each resistor?
(b) If the 8.0 Ω resistor were shorted out, what would be the new voltages across the remaining resistors?

21. (E15)
Resistors of values 2.0 Ω and 4.0 Ω are connected in parallel across
a 12 V d.c. supply.
(a)   What p.d. develops across each resistor?
(b)   What current flows through the 2.0 Ω resistor?
(c)   What current flows through the 4.0 Ω resistor?
(d)   What total current is delivered by the supply?
(e)   What must be the effective resistance of the circuit?

22. (E16)
(a)   What is the effective resistance of a 2.0 Ω and an 8.0 Ω resistor
when connected together
(i)   in series
(ii)   in parallel?
(b)   A 6.0 Ω resistor is connected in parallel with a 3.0 Ω resistor
and then they are connected in series with a resistor of value
x Ω. The effective resistance of this entire arrangement is
10.0 Ω. What must be the value of x?

23. (E18)
Millie is given a 24 W bulb. She noticed that the voltage stamped on
the bulb, was '12 V', but the current rating had been scraped off.
(a)   What would the current rating be?
(b)   Calculate the operating resistance of this bulb.

24. (E18)
A large electric fire is connected across a 250 V supply and a current
of 12 A flows.
(a)   Calculate the power rating of this appliance.
(b)   How much energy would be transformed when this fire is
switched on for one hour?

25. (E19)
Billy Blunder knows the colours of the three wires used in modern
flex, but he gets them all mixed up when he records them in his
physics notebook, as shown below.

Wire	Colour
earth	brown
neutral	green and yellow
live	blue

Rewrite the table in its correct form.

26. (E19)
A school technician has a supply of fuses with the following ratings:
1 A, 2 A, 5 A, 10 A and 13 A. Which fuse should he use to protect
the appliances listed below?
(a)   An h.t. power unit marked 600 V; 500 mA.
(b)   An electric fire marked 250 V; 2 kW.

(c) An electric bulb marked 12 V 36 W.
(d) An electronic desk calculator marked 115 V 1 A.
(e) An electric motor marked 250 V 3 kW.

27. (E20)

A copper rod is held at right angles to a uniform magnetic field. Both the rod and the field lines are horizontal.
(a) What happens when a current is passed through the rod?
(b) What rule helps to predict the direction of the force which acts upon the rod?
(c) Suggest ways of increasing the force on the copper rod.

28. (E20)

A pupil is supplied with a large horse-shoe magnet, a long piece of copper wire and a centre-zero galvanometer.
(a) How could he use this apparatus to demonstrate the conversion of mechanical energy into electrical energy?
(b) The direction of the induced electron current may be predicted using the left-hand generator rule. Give an outline of this rule.

29. (E23)

An oscilloscope is switched on and adjusted to give a small spot at the centre of the screen. What input should be applied to produce each of the following effects?
(i)   a single vertical line on the screen
(ii)   a single horizontal line on the screen
(iii) a spot to the right of the centre of the screen
(iv)   a spot in the lower left corner of the screen

30. (E23)

An oscilloscope with its time-base switched on has an a.c. signal applied to its Y-plates and a single (sine) wave appears on its screen. How would each of the following changes alter the original trace?
(i)   the time-base is slowed down
(ii)   the Y-gain is reduced
(iii) an a.c. supply of twice the peak voltage is used
(iv)   an a.c. supply of twice the frequency is used

UNIT R

Radioactivity

Content List of Topic Areas

Worked Examples

R1

The device shown in this circuit is called a Geiger–Müller tube and is used in the detection of radioactivity.

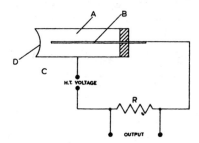

(a) (i)   What kind of radiations does this device detect?   (3)
     (ii) Name the parts of the tube labelled A, B, C and D.   (2)
     (iii) Describe briefly how this device detects radiation.   (2)

(b) If the voltage between B and C is varied from 0 to 500 V, the count-rate, measured by a counter unit connected across resistor R, varies as follows when a source is placed a fixed distance in front of D.

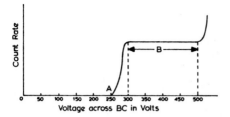

(i) Point A on the graph is referred to as the threshold voltage for the tube. What is meant by this? (1)

(ii) What is part B of the graph called and why should this particular tube be operated at around 400 V? (2)

R2
Alfie Ray, a scintillating young physicist, has succeeded in deflecting radiations from a radium source. The apparatus he used is shown below.

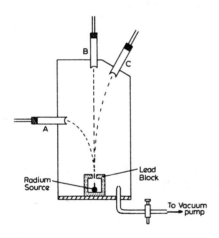

(a) Why is the radium source kept inside a lead block with only a small aperture? (1)

(b) Three detectors are used (A, B and C), each detecting one of the three types of radiation emitted by the radium source.
(i) What are the three radiations called? (3)
(ii) Which detector detects which radiation? (3)
(iii) Explain how radiations could be deflected as shown in the diagram. (2)

(c) Why must a vacuum pump be used at the start of this experiment? (1)

R3
The following graph indicates how the activity of a radioactive source decreases as time goes on. It is sometimes called a decay curve. The activity of the source is A when measurements are first taken.

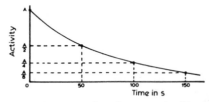

(a) What is meant by the term 'radioactive decay'?   (3)
(b) How long does it take for the activity of the source to decrease
     by 50%?   (2)
(c) What name is given to this time interval.   (1)
(d) What would be the activity after 250 s?   (3)
(e) Why is the count-rate from a radioactive source unlikely to
     reach zero in a practical experiment?   (1)

R4

At the end of his course in radioactivity, Mr Leadbetter draws up the
following chart and asks his pupils to write **'yes'** or **'no'** in each
of the boxes. The first row is filled in to illustrate the method. Find
out how much you know about alpha, beta and gamma radiations
by filling in all the vacant boxes.

Description or property	Alpha	Beta	Gamma
1. a high-speed electron	no	yes	no
2. a helium nucleus			
3. an electromagnetic radiation			
4. is positively charged			
5. is negatively charged			
6. is not charged			
7. is deflected by electric fields			
8. is deflected by magnetic fields			
9. is not deflected by electric and magnetic fields			
10. is absorbed by a thin sheet of paper			
11. is absorbed by a few cm of aluminium			
12. is absorbed by a very thick block of lead			
13. only penetrates a few cm through air			
14. is often symbolised by $^4_2$He			
15. is often symbolised by $_{-1}^{0}$e			
16. can be detected on a spark counter			
17. gives short, straight tracks in a cloud chamber			
18. is emitted from the nucleus of a radioactive substance			
19. mass number is reduced by its emission			
20. atomic number is increased by its emission			
21. atomic number is decreased by its emission			

                                                                       (10)

Solutions to Worked Examples

R1

(a) (i) This device detects nuclear radiations. If the end-window (D) is of a suitable thickness, the tube can detect all three nuclear radiations: alpha radiation (α), beta radiation (β), gamma radiation (γ).

(ii) A: Low pressure gas.
B: This is the anode and is usually a narrow metal rod.
C: This is the cathode and is usually a metal cylinder inside a glass wall.
D: This is the mica window through which the radiation enters the tube.

(iii) Radiation entering through the mica window ionises the inert gas. The ions are swept towards the charged electrodes (B and C) by the high voltage, and they cause a pulse of current in the circuit. A voltage pulse develops across R and is recorded on the counting-unit.

(b) (i) The threshold voltage is the voltage at which the tube just starts to respond to the incoming radiations.

(ii) B is called the plateau region of the graph. In this region, the count-rate is almost steady. The mid-plateau voltage is 400 V (see graph) and this is taken as the working voltage for this tube.

R2

(a) The radium source actually emits radiation in all directions. A lead block is used to make sure that radiation not passing through the aperture is absorbed by the lead. A controlled beam is thus obtained and the experiment is safer.

(b) (i) alpha (α), beta (β) and gamma (γ) radiations are emitted

(ii) A detects β-particles, B detects γ-rays and C detects α-particles.

(iii) A magnetic (or electric) field will deflect α and β particles, because they are charged, but they will not deflect γ-rays, which are not charged.

Magnetic field deflection
The magnetic field should be directed up out of the plane of the diagram to produce the deflections shown. Since β-particles are high speed electrons, they obey the **right-hand motor rule** which was discussed in question E20 of the electrical section. α-particles have positive charge and so they are deflected the opposite way.

Electric field deflection
The electric field should be directed across the diagram from left to right. If two parallel plates are used, the positive plate should be at the same side as detector A. The negatively charged β-particles will deflect towards the positive plate. The positively charged α-particles will deflect towards the negative plate.

(c)   A good vacuum is necessary to increase the range of the alpha-particles. At normal air pressure, the range of the alpha particles would be a few centimetres and they would not reach detector C before being absorbed.

R3
(a)   The atoms of a radioactive substance are unstable and they become more stable by emitting radiation (alpha, beta and/or gamma). Once an atom has become stable, it will emit no further radiation and so the sample becomes less active (emits fewer radiations) as time goes on. This process is called "radioactive decay".
(b)   50 seconds are required for the activity to decrease by 50% i.e. from A to $\frac{A}{2}$, or from $\frac{A}{2}$ to $\frac{A}{4}$, or from $\frac{A}{4}$ to $\frac{A}{8}$.
(c)   This time interval is called the **half-life** of the substance.
(d)   After 250 s (i.e. 100 s after the recorded time on the graph), the activity would be down to $\frac{A}{32}$. This is after 5 half-lives (250 = 5 x 50).
(e)   The count-rate in a real experiment will eventually reach background radiation level, unless, of course, a correction for this has been made before processing the results.

R4
1 NYN (given) 2 YNN 3 NNY 4 YNN 5 NYN 6 NNY 7 YYN
8 YYN 9 NNY 10 YNN 11 YYN 12 YYY 13 YNN 14 YNN
15 NYN 16 YNN 17 YNN 18 YYY 19 YNN 20 NYN 21 YNN

Practice Questions

1. (R1)
(a)   In a period of one second, one kilogram of radium undergoes 3.7×10^{13} disintegrations. If one curie (1 Ci) amounts to 3.7×10^{10} disintegrations per second, find the disintegration rate, in curies, of 0.1 g of radium.
(b) How does a Geiger–Müller tube detect such disintegrations?

2. (R2)
A radioactive source emits alpha, beta and gamma radiations. Draw a diagram to show how these radiations are affected by a magnetic field.

3. (R3)
The half-life of $^{220}_{86}Rn$ is 1 minute. If the activity of a sample of this substance is measured as 160 counts per minute, what would be the activity five minutes later?

4. (R3)

The activity of a radionuclide used in a biology experiment is recorded daily and corrected for background count. The results of the experiment are shown below:

corrected count rate (counts per second)	100	56	32	18	10	5.6
time (in days)	0	1	2	3	4	5

Plot these results on a graph and find the half-life of the radionuclide.

5. (R4)

Study the following nuclear equation.

$$^{234}_{90}X \longrightarrow {}^{234}_{91}Y + {}^{0}_{-1}e$$

(a) What does the symbol $^{234}_{90}X$ represent?

(b) What does the symbol $^{234}_{90}Y$ represent?

(c) What has been emitted in this reaction?

6. (R4)

Complete the nuclear reaction in which one alpha particle and gamma radiation is emitted.

$$^{238}_{92}U \longrightarrow {}^{(?)}_{(?)}Th + (?) + \gamma$$

Answers to practice questions

Wave Motion

1. (W1)   (a) transverse (b) 1.5 m (c) standing wave pattern

2. (W1)   (a) $v = f\lambda$ (m s^{-1}, Hz, m) (b) 20 m s^{-1} (c) 8 Hz
        (d) 1.25 m

3. (W2)   (a) (i) move end sideways, to and fro (ii) move end in
           and out
        (b) 2 m s^{-1}

4. (W3)   (a) 0.1 s (b) 2.5 mm (c) 30 cm s^{-1}

5. (W5)   (a) refraction
        (b) (i) decreases (ii) decreases (iii) stays same

6. (W5)   (a) 4 cm (b) 6 cm

7. (W5)   see solution to W5 (b) (i) for diagram

8. (W5)   (a) It should show plane waves growing in wavelength
        steadily from end X to Y. (b) no (c) yes

9. (W6)   (a) 30°, since angle of reflection is equal to angle of
        incidence
        (b) (i) none (ii) none (iii) none

10. (W6)   (a) a concave reflector. See solution to W6 (a) (i)
        (b) (i) diffraction (ii) see solution to W6 (a) (iii)/(iv)

11. (W6)   (a) crest meets trough causing destructive interference
        (b) crests arrive together and then troughs arrive
        together causing constructive interference

12. (W7)   (a) the wider the gap, the less the diffraction (b) low
        frequency (c) no

13. (W7)   diagram is similar to W7 diag. (iv)

14. (W8)   (a) red (b) violet (c) red (d) yellow (e) white

15. (W9)   (a) see solution to W9 Box D (b) see solution to W9
        Box E (c) see solution to W9 Box B with direction
        of rays reversed

16. (W10)   (a) see diagram in W10
        (b) (i) violet (ii) red
        (c) infra-red radiation

17. (W11)   (a) metals (b) paraffin (c) about 3.0 cm (d) microwave
        transmitter unit, 2 large sheets of metal and 1 strip
        of metal, to form 2 gaps, microwave aerial probe,
        display unit.

18. (W12)   gamma, X-rays, u.v., visible light, infra-red, radio
        waves (b) 3.0×10^8 m s^{-1} (c) 1.0×10^{10} Hz
        (d) 3.0 m

19. (W13)   (a) Sound waves travel much slower than light waves.
           (b) 495 m
20. (W14)   (a) longitudinal (b) 330 m s⁻¹ (c) 0.5 m
           (d) (i) fewer waves on screen (ii) amplitude of trace
               decreases
21. (W15)   see solution to W15 (a) (i) for diagram
22. (W15)   signal generator, two loudspeakers, connecting wire,
           microphone, display unit (e.g. oscilloscope)

Mechanics

1. (M1)   (a) single stationary line (b) two stationary lines at
          180° (c) three stationary lines at 120° (d) same
          as (a) (e) same as (a)
2. (M2)   401 dots
3. (M2)   (a) $\frac{1}{40}$ s (b) 40 Hz
4. (M3)   (a) 9 km
          (b) (i) 2 m s⁻¹ (ii) 15 km at 37° N of E (iii) 1.43 m s⁻¹
          at 37° N of E
5. (M4)   (a) 8 m s⁻² (b) 16 m s⁻¹ (c) 12 m s⁻¹
6. (M4)   (a) 5 m s⁻¹ (b) 2 m s⁻²
7. (M5)   (a) 90 m s⁻¹ (b) 405 m
8. (M6)   Gradients of A, B and C are 0, +ve and −ve respectively
          (see graph in M6).
9. (M6)   (a) displacement (b) acceleration
10. (M6)   (a) 10 m s⁻¹ (b) 100 m
          (c) (i) 5 m s⁻¹ (ii) 5 m s⁻¹ (iii) 6.7 m s⁻¹
11. (M7)   5.0 m s⁻²
12. (M8)   (a) see diag. M8 (b) no difference in the spacing
13. (M9)   read solution to M11 (a)
14. (M9)   Trolley will move at constant velocity.
          Trolley would slow down due to friction.
15. (M10)   (b) 2 units (c) 3 units (d) 2 units (e) 3 units
          (f) 6 units
16. (M11)   (a) It is 1 unit. (b) It is more than 1 unit (c) It is less
          than 1 unit.
17. (M11)   (a) N, kg, m s⁻² (b) 4.0 m s⁻² (c) 4.0 kg (d) 800 N
          (weight)
18. (M12)   (a) 10 000 N (b) −6000 N (c) 480 m
19. (M13)   (a) −1000 N (b) 0.25 m s⁻² (c) 600 N
20. (M14)   (a) 2.0 s (b) 16 m
21. (M15)   (a) (i) 1800 N (ii) 0.20 N (iii) 12 000 N
          (b) 900 N, 0.10 N, 6000 N (c) all same.
22. (M15)   'To every action, there is an equal and opposite
          reaction'.
          (a) The force of the boot on the ball is equal and
             opposite to the force of the ball on the boot.
          (b) The force of the rocket on the exhaust gases is
             equal and opposite to the force of those gases on
             the rocket.

23. (M16) (a) 6000 kg m s^{-1} (b) 6000 kg m s^{-1} (c) 3.0 m s^{-1}
(d) inelastic collision
24. (M16) (a) 1.0 m s^{-1} to the right (b) 5000 J
25. (M17) 5 m s^{-1}
26. (M18) (a) 0 kg m s^{-1} (b) 4.0 kg (c) 28 J
27. (M18) 0.48 m s^{-1}
28. (M19) (a) equal and opposite (b) equal and opposite
(c) Pupils have same size of momentum. The one
with larger mass has slower speed. (d) zero.
29. (M20) (a) (i) 1200 J (ii) 900 J
(b) 6000 J
30. (M20) 2000 W
31. (M21) (a) 2500 W (b) 5000 N

Heat and the Gas Laws
1. (H1) 13.2 kJ
2. (H2) (a) specific heat capacity (b) E energy; c specific heat
capacity; m mass of substance; ΔT change in
temperature; (c) read solution to H2
3. (H3) 625 J kg^{-1} K^{-1}
4. (H3) 84 kJ
5. (H4) (a) 25 J (b) 7.5 kJ
6. (H4) 4.1°C
7. (H6) (a) 8.4. kJ (b) 8.4 kJ (c) 700 J kg^{-1} K^{-1}
8. (H7) 5.01 x 10^4 J
9. (H7,9) (i) 3.39 x 10^5 J (ii) 6.3 x 10^4 J (iii) 5.01 x 10^4 J
10. (H8,9) 3.62 x 10^6 J
11. (H10) (a) temperature of gas (b) $\dfrac{pV}{T_K}$ = constant (c) 3.3 units
(d) 10 units
12. (H11) 0.16 m^3
13. (H12) (a) read solution to H14 (b) 1.2 x 10^5 Pa
14. (H14) new pressure = 6 x old pressure
15. (H13,14) 0.05 m^3
16. (H13,14) 0.78 m^3

Electricity
1. (E1) (a) positive (b) repel (c) repel (d) negative
2. (E2) read solution to E2
3. (E3) (a) leaves rise (b) leaves fall
4. (E3) (a) surplus (b) shortage
5. (E4) see solution to E4 (b) (i), (ii) and (iii)
6. (E6) (a) 12 C
7. (E6) (a) 2 A (b) 1 A
8. (E7) (a) 240 V (b) 115 V
9. (E7) 2 A
10. (E8) (a) current is same through each bulb (b) Bulbs would
get brighter. Resistance of circuit is less and current
is increased.
11. (E9) (a) 2 A through each bulb (b) Bulbs stay at same brightness if supply voltage is constant.

12. **(E9)** (a) 3.0 A (b) 0.8 A (c) 2.5 A (d) 1.5 A
13. **(E9)** A middle B middle C bottom D bottom
14. **(E10)** (a) (i) in series (ii) in parallel
    (b) 6 cells
15. **(E10)** (i) must be low (ii) must be high
16. **(E11)** (a) 4 ohms (b) x is 0.5 A   y is 2.5 A (c) 0.75 A
17. **(E12)** (a) 80 ohms (b) 15 A (c) 12 V (d) 200 ohms
18. **(E13)** (a) 2 A
    (b) (i) 4 V (ii) 8 V
19. **(E13)** (a) 2400 ohms (b) 12 V (c) 0.1 A
20. **(E14)** (a) 1 V, 3 V, 4 V and 8 V respectively (b) 2 V, 6 V
    and 8 V respectively
21. **(E15)** (a) 12 V (b) 6 A (c) 3 A (d) 9 A (e) 1.33 ohms
22. **(E16)** (a) (i) 10.0 ohms (ii) 1.6 ohms (b) 8 ohms
23. **(E18)** (a) 2 A (b) 6 ohms
24. **(E18)** (a) 3 kW (b) 1.08×10^7 J
25. **(E19)** earth is green and yellow, neutral is blue, live is brown
26. **(E19)** (a) 1 A (b) 10 A (c) 5 A (d) 2 A (e) 13 A
27. **(E20)** (a) rod is moved vertically (b) right-hand motor rule
    (c) increase magnetic field, increase current through
    rod
28. **(E20)** (a) Attach ends of wire to galvanometer and then move
    wire at right angles to field of magnet. (b) see
    solution to E20 (d) (ii)
29. **(E23)** (i) Apply alternating voltage to Y-plates. (ii) Apply
    alternating voltage to X-plates, or switch on fast
    time base. (iii) see solution to E23 (a) (i).
    (iv) Apply d.c. signals to both plates with left and
    bottom plates both positive.
30. **(E23)** (i) more than one sine wave on the screen (ii) amplitude
    of trace reduces (iii) amplitude (peak) would
    double (iv) two waves would appear on the screen

Radioactivity
1. **(R1)** (a) 0.1 Ci (b) read solution to R1
2. **(R2)** read solution to R2
3. **(R3)** 5 counts per minute
4. **(R3)** 1.2 days
5. **(R4)** (a) radioactive beta emitter of mass number 234 and
    atomic number 90 (b) substance whose mass
    number is 234 and whose atomic number is 91
    (c) beta particle
6. **(R4)** $^{238}_{92}U \longrightarrow {}^{234}_{90}Th + {}^{4}_{2}He + \gamma$

INDEX

Index to worked examples

116